1976

INTERNATIONAL SERIES OF MONOGRAPHS IN

ANALYTICAL CHEMISTRY

General Editors: R. BELCHER and H. FREISER

Volume 34

ENZYMATIC METHODS
OF ANALYSIS

ENZYMATIC METHODS
OF ANALYSIS

by

GEORGE G. GUILBAULT

*Department of Chemistry, Louisiana
State University, New Orleans, La.*

PERGAMON PRESS

*Oxford · New York · Toronto
Sydney · Braunschweig*

Pergamon Press Ltd., Headington Hill Hall, Oxford
Pergamon Press Inc., Maxwell House, Fairview Park, Elmsford, New York 10523
Pergamon of Canada Ltd., 207 Queen's Quay West, Toronto 1
Pergamon Press (Aust.) Pty. Ltd., 19a Boundary Street, Rushcutters Bay, N.S.W. 2011,
Australia
Vieweg & Sohn GmbH, Burgplatz 1, Braunschweig

First edition 1970
Reprinted 1973
Library of Congress Catalog Card No. 72-100363

Printed in Great Britain by A. Wheaton & Co., Exeter

ISBN 0 08 006989 4

CONTENTS

v

CONTENTS   xi

PREFACE

Enzymes possess a great potential usefulness in
analytical chemistry. The specificity of enzymes
can solve the primary problem of most analytical
chemists, the analysis of one substance in the
presence of many similar compounds that interfere
in the analysis. The sensitivity of enzymes allows
the determination of as little as 10^{-10}g of material.

Enzyme catalyzed reactions have been used for
analytical purposes for many years for the determination of substrates, activators, inhibitors, and
also of enzymes themselves. Until recently, however,
the disadvantages associated with the use of enzymes
have seriously limited their usefulness. Frequently
cited objections to the use of enzymes for analytical
purposes have been their unavailability, instability,
poor precision, and the tediousness of analyses.
While these objections were valid earlier, numerous
enzymes are now available in purified form, with
high specific activity, at reasonable prices. The
instability of enzymes is, of course, always a
potential hazard; yet, if this instability is
recognized and reasonable precautions are taken,
this difficulty can be minimized. The poor precision,
slowness and labor that have made enzyme-catalyzed
reactions unappealing as a means of analysis may be
more a consequence of the methods and techniques
than the fault of the enzymes.

With the new, more sensitive techniques available
for assay of enzymes, the advent of the immobilized
enzyme which has alleviated the problems of cost and
supply, and the progress that has been made in automation of enzyme systems for rapid, accurate analysis,
enzymes are becoming common, acceptable reagents by
analysts. This is clearly indicated by the large
increase in the number of scientific papers published
in this area in the past few years.

My aim in writing this book is to cover all the
aspects of modern enzymatic analysis. Two introduc-
tory chapters are devoted to general considerations
of enzymes as reagents and methods of analysis of
enzymatic reactions. The next four chapters deal
with methods for the assay of specific enzymes, and
substrate, activator and inhibitor analysis using
enzymes. In the last two chapters the immobilization
of enzymes and the automation of enzymatic reactions
are discussed. In addition, a listing of all commercially available enzymes is given in an appendix.
It is hoped that the information presented will prove
interesting and stimulating to all individuals en-
gaged in research and development.

I wish to express my appreciation to those who
have examined and critically reviewed the manuscript
or portions of it: Professor H. Freiser, University
of Arizona; Professor A. Townshend and Professor
R. Belcher, University of Birmingham; Dr. R. Phillips,
Turner; Professor P. Hicks, University of Wisconsin;
Professor H. Pardue, Purdue University; Professor B.
Kratochvil, University of Alberta; Professor E. C.
Toren, Duke University; and Dr. J. Levine, Technicon,
Inc. I would like to thank Mrs. Mercedes Weiser for
typing the entire book in final form for direct repro-
duction.

PREFACE   xv

I am especially grateful to my wife, Palma Covington
Guilbault, for her constant encouragement throughout
the years, and dedicate this book to her.

                              George G. Guilbault

New Orleans, Louisiana
January, 1969

To Pal

CHAPTER 1

GENERAL CONSIDERATIONS

A. PRINCIPLES OF ENZYMATIC ANALYSIS

Enzymes are biological catalysts which enable the
many complex chemical reactions, upon which depends
the very existence of life as we know it, to take
place at ordinary temperatures. Because enzymes work
in complex living systems one of their outstanding
properties is specificity. An enzyme is capable of
catalyzing a particular reaction of a particular substrate even though other isomers of that substrate
and other compounds of similar structure may be present.

An example of the selectivity of enzymes with respect to a particular substrate is illustrated by luciferase, which catalyzes the oxidation of luciferin(I)
to oxyluciferin.[1] A complete study of many compounds
similar in structure to luciferin, showed that the
catalytic oxidation resulting in the production of
the green fluorescence occurs only with luciferin.
Addition of another hydroxy group or substitution

of an amino group in the luciferin molecule alters
the enzymatic reaction and no green luminescence is
observed.

1

Glucose oxidase, which catalyzes the oxidation of β-D-glucose to gluconic acid is still more selective. In a study of 60 oxidizable sugars and their derivatives, workers[2] have found that only β-D-glucose, 2-deoxy-D-glucose and 6-deoxy-fluoro-D-glucose are oxidized at an appreciable rate. The anomer α-D-glucose is oxidized catalytically at a rate less than 0.6% as rapidly as the β-anomer.[2]

Enzymes also exhibit selectivity with respect to a particular reaction. If one attempted to determine glucose by oxidation in an uncatalyzed way, by heating a glucose solution with an oxidizing agent like ceric perchlorate, many side reactions occur to yield products in addition to gluconic acid. With the enzyme glucose oxidase, however, catalysis is so effective at room temperature and pH 7 that the rates of the other thermodynamically possible reactions are negligible.

This selectivity of enzymes, and their ability to catalyze reactions of substrates at low concentrations, is of great use in chemical analysis. Enzyme-catalyzed reactions have been used for analytical purposes for a long time in the determination of substrates, activators, inhibitors, and also of enzymes themselves. Osann[3] used peroxidase for an assay of peroxide in 1845, and enzymic methods were accepted techniques for the analysis of carbohydrates in the 19th century. In the early 1940's procedures based on the photometric measurement of the reduced coenzymes, nicotinamide adenine dinucleotide (NAD) and nicotinamide adenine dinucleotide phosphate (NADP) were described by Warburg.[4] This, and the development recently of new electrochemical and fluorometric methods, has made enzymic methods of analysis an accepted analytical technique.

B.   PROPERTIES OF ENZYMES

The basic equations for the reaction of an enzyme and its substrate were developed by Michaelis and

Menten. In the mechanism for the reaction, a substrate,
S, combines with the enzyme, E, to form an intermediate
complex, $[ES]$, which subsequently breaks down into
products, P, and liberates the enzyme. The enzyme is

$$E + S \underset{k_2}{\overset{k_1}{\rightleftharpoons}} [ES] \xrightarrow{k_3} E + P$$

a true catalyst, since it effects the transformation
of the substrate to products, yet is not consumed in
the reaction.

The equilibrium constant for the formation of the
complex, $[ES]$, is called the Michaelis constant, K_m,
which is defined as $(k_2 + k_3)/k_1$. The initial rate
of reaction, V_o, is then some function of the enzyme
and substrate concentration (equation 1):

$$V_o = V_{max} [S]_o / (K_m + [S]_o) \qquad (1)$$

where $[S]_o$ is the initial substrate concentration and
V_{max} is the maximum rate of reaction.

At a fixed enzyme concentration the rate increases
with substrate concentration,until a non-rate-limiting
excess of substrate is reached (Fig. 1), after which
addition of more substrate causes no increase in rate.
When $V = V_{max}/2$, $[S] = K_m$, and K_m can be determined.
The reciprocal of equation (1) is:

$$\frac{1}{V_o} = \frac{K_m}{V_{max} [S]} + \frac{1}{V_{max}} \qquad (2)$$

and a plot of $1/V_o$ vs. $1/[S]$ yields a straight line
(Fig. 2) with an intercept of $\frac{1}{V_{max}}$ and a slope of
K_m/V_{max}. This plot, ascribed to Lineweaver and Burk[5],
is the most common method for determining K_m.

The Michaelis constant, being a constant for the dis-
proportionation of the enzyme substrate complex, is a
good indication of the quantitativeness of an enzyme
reaction. The smaller the value of K_m, the more

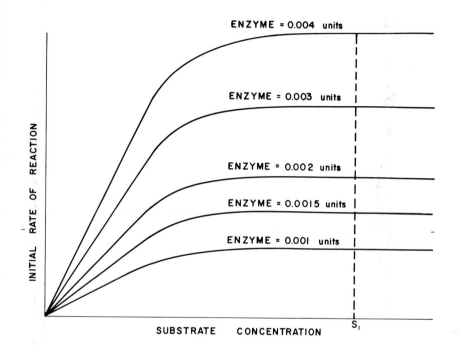

FIG. 1

Plot of initial rate vs. substrate concentration
at various enzyme concentrations. S_1 = non-rate-
limiting substrate concentration.

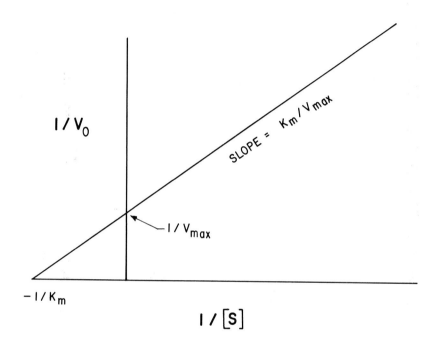

FIG. 2
Lineweaver-Burk Plot of $1/V_0$ vs. $1/[S]$

complete the enzyme-substrate reaction will be.

Another indication of the rate of an enzymic reaction is k_3, the constant for the conversion of enzyme-substrate complex into products. It is linearly related to V, via the enzyme concentration:

$$V_{max} = k_3 \left[E \right] \qquad (3)$$

Thus knowing the enzyme concentration and the V_{max} one can calculate k_3.

C. FACTORS INFLUENCING THE ENZYME REACTION RATE

The chief factors which determine the initial velocity of an enzymic reaction are the enzyme concentration, the substrate concentration, pH, temperature, activators, inhibitors and ionic strength.

1. Effect of Enzyme Concentration

As predicted from the Michaelis equation, the initial rate of an enzymic reaction is proportional to the initial enzyme concentration, $\left[E \right]_o$. This dependence is illustrated in Fig. 1.

$$V_o = \frac{k_3 \left[E \right]_o \left[S \right]_o}{K_m + \left[S \right]_o} \qquad (4)$$

Theoretically an increase in rate should be observed for each increase in enzyme concentration ad infinitum. It is sometimes found that there is a falling off from linearity at very high enzyme concentration. This does not indicate a true decrease in the activity of enzyme but represents a limitation in the technique of measurement. Thus from a measurement of the initial rate and a calibration plot of the rate vs. enzyme concentration one can easily calculate the concentration of this biochemical catalyst.

Some enzymes have been observed to give non-linear plots of rate vs. enzyme concentration usually showing a curvature towards the horizontal axis. Proteinase, for example, acting on proteins, fits the equation $V_o = k [E]^{\frac{1}{2}}$, called the Schutz law.[6,7] A number of proteinase preparations acting on hemoglobin or gluten fit the equation, $V_o = k [E]^{2/3}$, and some authors suggest the use of this equation in the determination of proteolytic activity.[8] Roy[9] found that the velocity of the arylsulphatase A reaction followed the form, $V_o = k [E]^{3/2}$ and offered a theoretical explanation in terms of dimerization of the enzyme. In all these cases, however, deviations from linearity are probably due in some measure to the presence of activators or inhibitors in the enzyme preparation. In the vast majority of cases an exact proportionality between initial velocity and enzyme concentration has been found and in most kinetic studies this proportionality is assumed.

2. Effect of Substrate Concentration

The concentration of substrate is one of the most important factors affecting the rate of an enzymic reaction. A plot of the initial velocity versus the substrate concentration is a section of a rectangular hyperbola as indicated in Fig. 1. As predicted from the Michaelis equation (equation 4), when $[S] \ll K_m$, $V_o = \dfrac{k_3 [E] [S]}{K_m}_o$ and the rate will be proportional to the substrate concentration. As $[S] \gg K_m$, $V_o = k_3 [E]$ ($= V_{max}$) and the rate becomes independent of substrate concentration. At the point where $[S] = K_m$, $V = \frac{1}{2} V_{max}$. For an analytical method for determining substrate concentration, therefore, $[S]$ must be $\ll K_m$, so that the enzymic reaction becomes first order with respect to substrate. For optimum precision, $[S]$ should be $< 0.1 K_m$.

Frequently a decrease in the rate of an enzyme reaction is observed at high substrate concentration. This

substrate   inhibition is not predictable from the
Michaelis expression, and may be due to a number of
different causes.   For example ,in the hypoxanthine-
xanthine oxidase reaction, monitored by the reduction
of the highly colored methylene blue, the inhibition
by substrate is due to competition with methylene blue.
In other cases this inhibition is due to the formation
of ineffective complexes with two or more substrate
molecules combined with one active site.[2]   This is
observed with many enzymes which have two or more groups,
each combining with a particular part of the substrate
molecule.   In the "effective" complex one substrate
molecule is combined with all these groups.   If some
of these groups are blocked with other molecules an
"ineffective" complex can be formed in which a substrate combines with only one group on the enzyme.   The
chance of an ineffective complex forming increases at
high substrate concentration, where the substrate molecules tend to crowd onto the enzyme.

3.   Effect of Activators

   The rate of some enzymic reactions can be greatly
increased by the addition of very small amounts of
certain substances called activators.   An enzymic
activator is a substance which is required for an
enzyme to be an active catalyst:

$$E \text{ (inactive)} + \text{activator} \rightleftharpoons E \text{ (active)}$$

$$E \text{ (active)} + \text{Substrate} \rightleftharpoons \text{Products}$$

Some activators merely increase the efficiency of an
already functioning enzyme, i.e. Mg^{++} activation of
alkaline phosphatase.   The activity of the enzyme will
increase until enough activator has been added to activate the enzyme fully.   The initial rate of the enzyme
reaction is proportional to the activator concentration
at low concentrations, thus providing a method for its
determination (Fig. 3).   At high concentrations, the

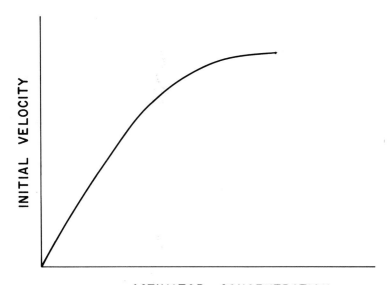

FIG. 3

Variation of the initial velocity of an enzyme reaction with increasing concentration of an activator.

rate becomes independent of the amount of activator.

Many workers have described the activation of the enzyme isocitric dehydrogenase by metal ions, such as Mn^{2+} and Mg^{2+}.[10-12] As little as 5 parts per billion of Mn^{2+} can be determined based on this activation.

For the analysis of substrates in enzyme-activated systems, an excess, non-rate-limiting concentration of activator is used. Under these conditions the rate becomes pseudo first order with respect to substrate concentration.

4. Effect of Inhibitors

An inhibitor is a compound that causes a decrease in the rate of a catalytic reaction, either by reacting with the catalyst to form a catalyst-inhibitor complex or by reacting with one of the reactants. Enzymic inhibitors are either reversible or irreversible. In reversible inhibition, the enzyme can recover its activity when the inhibitor is removed; in the case of an irreversible inhibitor it does not. With an irreversible inhibitor the inhibition increases progressively with increasing inhibitor concentration and becomes complete if enough inhibitor is present to combine with all the enzyme. With a reversible inhibitor inhibition is progressive, but quickly reaches an equilibrium value which depends upon the inhibitor concentration. The action of the "nerve gases", Sarin and Tabun, on cholinesterase is an example of irreversible inhibition; eserine, however, is a reversible inhibitor of cholinesterase.

a. Reversible Inhibition. In competitive inhibition both substrate and inhibitor compete equally for the active site of the enzyme. A fully competitive type of reversible inhibition can be represented by the following reactions:

$$E + S \underset{k_2}{\overset{k_1}{\rightleftharpoons}} [ES]$$

$$E + I \underset{k_6}{\overset{k_5}{\rightleftharpoons}} [EI]$$

$$[ES] \underset{k_4}{\overset{k_3}{\rightleftharpoons}} E + P$$

Applying steady-state kinetics and solving for V_o we obtain

$$V_o = \frac{k_3 \, [E]}{1 + \frac{k_2 + k_3}{k_1 [S]} (1 + \frac{k_5 [I]}{k_6})}$$

Writing K_m for $(k_2 + k_3)/k_1$ and K_i for k_6/k_5 we obtain

$$V_o = \frac{V_{max}}{1 + \frac{K_m}{[S]} (1 + \frac{[I]}{K_i})}$$

The effect of the competitive inhibitor is to produce an apparent increase in K_m by the factor $1 + [I]/K_i$. Thus the apparent K_m will increase without limit as $[I]$ increases. At any inhibitor concentration, the limiting velocity with excess of substrate is always equal to V_{max}, the maximum velocity of the uninhibited reaction. Typical plots of V vs. [S] and 1/V vs. 1/[S] for competitive inhibition are indicated in Fig. 4.

    b. Irreversible Inhibition. Equations for non-competitive, irreversible inhibition are

$$E + S \rightleftharpoons [ES]$$

$$E + I \rightleftharpoons [EI]$$

$$[EI] + S \rightleftharpoons [EIS]$$

$$[ES] + I \rightleftharpoons [EIS]$$

$$[ES] \longrightarrow E + P$$

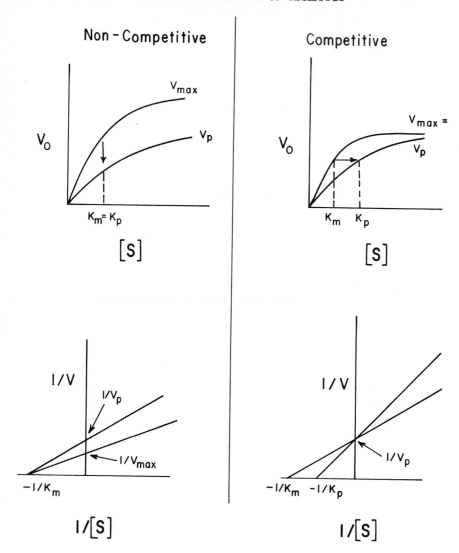

FIG. 4

Plots of V_O vs. [S] and 1/V vs. [S] for non-competitive and competitive inhibition. V_p is the maximum velocity in excess substrate. K_p is the substrate concentration giving half the maximum velocity.

Assuming the complex [EIS] does not break down and the velocity is entirely that of the breakdown of [ES], an expression for the initial velocity, V_o is:

$$V_o = \left(\frac{V_{max}}{1 + \frac{[I]}{K_i}} \right) / \left(1 + \frac{K_m}{[S]} \right)$$

The initial rate of an enzyme reaction, then, will decrease with increasing inhibitor concentration, linearly at low inhibitor concentration, then asymptotically reaching zero (Figure 5). Generally, kinetic methods are extremely sensitive for determining substances which are catalytic inhibitors. For example, as little as 10^{-10} g per ml. of organophosphorus compounds can be determined by their inhibition of the enzyme cholinesterase, which catalyzes the hydrolysis of choline esters.[13] Moreover, a great deal of specificity is built into inhibitors, thus providing an additional advantage in analysis. A specific method for the determination of nanogram quantities of fluoride in the presence of phosphate was described by Linde[14] and McGaughey and Stowall.[15] Fluoride inhibits the enzyme liver esterase, but phosphate is a weak inhibitor. This provides one of the few direct methods for fluoride in the presence of large amounts of phosphate.

5. Effect of Temperature

The overall enzyme reaction consists of three successive stages: the formation of enzyme-substrate complex, conversion of this to an enzyme-product complex and dissociation to products and free enzyme. The total effect of temperature on the reaction will be the resultant of the separate effects of these individual steps. The heat, free energy and entropy of activation, and the heat, free energy and entropy of the process for each of these three stages will contribute

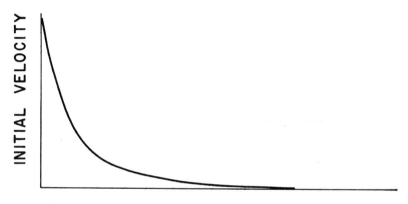

INHIBITOR   CONCENTRATION

FIG. 5

Variation of the initial velocity of an
enzyme reaction with increasing concentration
of an inhibitor.

to the overall thermodynamic parameters observed. A
classical derivation yields:

$$\frac{d \log k}{dt} = \frac{\Delta H^{\ddagger} + RT}{2.303 \ RT^2} = \frac{E^*}{2.303 \ RT^2}$$

A plot of log k vs. 1/T will give a straight line with
a slope of $-(\Delta H + RT)/2.303R$. It should be noted that
this plot gives E^* and not $\Delta H^{\ddagger}$.

The effect of temperature on enzyme reactions is
usually given in terms of the temperature coefficient,
Q_{10}, which is the factor by which the rate of reaction
is increased by raising the temperature $10^{\circ}C$.

$$E = \frac{2.303 \ RT^2 \ \log Q_{10}}{10}$$

Generally the temperature coefficient of an enzymic
reaction lies between 1 and 2; hence an increase in
temperature will increase the rate of the reaction,
and a decrease in temperature will decrease the rate.
Since a $10^{\circ}C$ rise in temperature will approximately
double the reaction rate in many cases, a strict control over the temperature must be maintained during
any kinetic experiment. A control to within $\pm 0.1^{\circ}C$
is generally sufficient to obtain good reproducible
results.

6. Effect of pH

Most enzymes are active over only a limited range
of pH and in most cases a definite optimum pH is observed. This pH optimum might be due to a number of
effects:

(1) an effect of pH on the stability of the enzyme,
which may become irreversibly destroyed on one or both
sides of the optimum pH,

(2) an effect on the V_{max} itself,

(3) an effect on the affinity of the enzyme, the

fall on either side of the optimum being due to a
decreased saturation of the enzyme with substrate, due
to a decreased affinity, and

(4)   an effect of pH on the indicator reaction, if
one is used to monitor the progress of the enzymic
reaction by a coupled reaction sequence. These effects
may occur in combination, and can easily be distin-
guished experimentally.

In the development of an analytical procedure the
pH dependence of the total enzyme system, including
the indicator reaction, is determined experimentally,
and the optimum pH is used for analysis. In many cases
a compromise must be made, if the optimum pH of each
reaction in a coupled sequence is different. For
example, in the determination of sucrose, a three step
enzymic reaction sequence is used. In the first step
sucrose is cleaved to glucose by the enzyme invertase
at a pH of 4.6. Glucose is then converted to peroxide
by glucose oxidase at pH 7-8.

$$\text{Sucrose} \xrightarrow[\text{pH 4.6}]{\text{Invertase}} \text{Glucose}$$

$$\text{Glucose} \xrightarrow[\text{pH 7-8}]{\text{Glucose Oxidase}} H_2O_2$$

$$H_2O_2 + \text{Indicator} \xrightarrow[\text{pH 10}]{\text{Peroxidase}} \text{Fluorescence}$$

In the final step, the peroxide produced is measured in
an indicator reaction which proceeds best at a pH 10.
A compromise pH of 8.5 allows the latter 2 reactions to
proceed but not the first, which does not go above pH 6.
At this pH of 6 or below, however, the latter reactions
do not work. Therefore, one must use a pH of 4.6 to
convert all the sucrose to glucose. Then the pH is
raised to pH 8.5 and the glucose produced is measured
by the latter reactions.

7. Effect of Ionic Strength

The presence of foreign salts can affect the rate of reaction, either by shifting the equilibrium of formation of the activated complex or by combining with reactants. The first, called a salt effect of the first kind, can be calculated from the basic equation

$$\log K = \log K_o + Z_A Z_B \sqrt{\mu} + (k_A + k_B - k^*) \mu$$

where K is the rate constant of the reaction, K_o the rate constant without foreign substances, Z_A and Z_B the charges on the reacting substances A and B, μ the ionic strength and k_A, k_B and k^* are empirical coefficients for A, B and activated complex, respectively.

In the salt effect of the second type, the foreign substances may serve to tie up the effective concentration of one of the reactants via the formation of a complex ion or a precipitate, by shifting the ionization equilibrium of weak acid or weak base reactants, or some similar effect. Hence, to achieve reproducible results, one must carefully eliminate harmful foreign ions and control the ionic strength of the medium. In an ionic reaction, the rate will vary with the dielectric constant of the solvent used.[16, 17]

D. DETERMINATION OF CONCENTRATIONS

The concentration of a material participating in an enzyme reaction can be calculated in one of two ways: by measuring the total change that occurs by chemical, physical or enzymatic analysis of the product or unreacted starting material; or from the rate of the enzyme reaction which depends on the concentration of the substrate, coenzyme, activator or inhibitor as discussed above.

1. Total Change or Equilibrium Method

In the first method, large amounts of enzyme and small

amounts of substrate are used to ensure a relatively
rapid reaction. The reaction is allowed to go to
completion, and the amount of substrate, S, in the
sample can be calculated from the amount of product,
P, formed: Substrate $\xrightarrow{\text{Enzyme}}$ Product. P must be
in some way chemically and/or physically distinguish-
able from S. For example, ethanol can be determined
by the enzymic reaction using alcohol dehydrogenase
in conjunction with the coenzyme nicotinamide adenine
dinucleotide, NAD.

Ethanol + NAD $\xrightarrow{\text{Alcohol Dehydrogenase}}$ Acetaldehyde + NADH.

The reduced form of NAD, NADH, is formed which has a
strong absorbance at 340 mμ, where NAD does not absorb.
The total amount of NADH produced is therefore a
measure of the amount of ethanol present. Likewise in
the determination of uric acid with the enzyme uricase,
peroxide is produced:

Uric $\xrightarrow{\text{Uricase}}$ H_2O_2

Uric acid has a strong absorbance at 290 mμ where the
products do not absorb. The total uric acid present
can therefore be determined by noting the total change
in the absorbance at 290 mμ.

Alternatively, a coupled reaction can be used to
indicate how much substrate has been decomposed. In
the determination of glucose an enzyme reaction using
glucose oxidase yields hydrogen peroxide. The extent
of reaction could be determined by monitoring the
uptake of oxygen using an oxygen sensitive electrode,
or, more easily, with the aid of an indicator reaction
which yields a colored dye from a colorless leuco-dye
(i.e., o-dianisidine).

Enzyme Reaction:

$$\text{Glucose} + H_2O + O_2 \xrightarrow{\text{Glucose Oxidase}} \text{Gluconic Acid} + H_2O_2$$

Indicator Reaction:

$$H_2O_2 + \text{leuco-dye} \xrightarrow{\text{Peroxidase}} H_2O + \text{dye}$$
$$\text{(colorless)} \qquad\qquad\qquad \text{(colored)}$$

The total intensity of color of the dye produced
is a measure of the concentration of glucose present.

2. Kinetic Method

In the second method, the kinetic method, the
initial rate of reaction, V_o, is measured in one of
the many conventional ways, by following either the
disappearance of substrate or the production of product. The rate is a function of the concentration
of substrate (S), enzyme (E), inhibitor (I) and activator (A). For example, the concentration of
glucose can be determined by measuring the initial
rate of production of the colored-dye in the example
given above.

Since the enzyme is a catalyst, and as such
affects the rate, and not the equilibrium, of a
reaction, its activity must be measured by a kinetic
(or rate) method, or by a direct titration of the
active site.[18] Likewise, activators or inhibitors
that affect the enzyme's catalytic ability can be
measured only by a rate change. The substrate can,
however, be measured either via a total change or
a kinetic method. The former frees the technician
from continuous measurements; rate methods, however,
are faster because the initial reaction can be
measured, without waiting for the reaction to go to
completion. The accuracy and precision of both
methods are comparable[19], and it is no longer

true that equilibrium methods are more reliable than
rate methods. The rate of reaction is affected by
conditions of pH, temperature and ionic strength,
however, and all these factors must be carefully controlled for good results. Recent work by Guilbault[20, 21] and Pardue[22] has indicated that with
reasonable care, precision and accuracies of better
than 1% can be obtained. Furthermore, some of the
difficulties encountered because of side reactions
are eliminated in rate methods and greater sensiti-
vities can be obtained in many cases. With the
automated equipment now available for performing
rate methods, such techniques will probably be the
ones of choice in the future.

There are several possible methods to calculate
the rate of an enzymic reaction: 1) the initial
slope method; 2) fixed concentration or variable
time method; and 3) fixed time method. These methods
and their automation are discussed in Chapter 8,
p. 266 et seq.

E.   HANDLING BIOCHEMICAL REAGENTS

The enzymes used as analytical reagents in enzymic
analysis are relatively fragile substances, which
have a tendency to undergo inactivation or denaturation if not properly handled. The first considera-
tion should also be given to the proper handling of
the enzymes, so as to avoid inactivation.

Generally, high temperatures and acid or alkaline
solutions are to be avoided. Most enzymes are
inactivated above 35-40°C (body temperature) and
in solutions of pH less than 5 or greater than 9.
In adjusting the pH of an enzyme solution, one must
be careful not to create a zone of destruction
around a drop of reagent added, which would tend to

inactivate some of the enzyme in solution. Solutions
of enzymes should be well stirred and acids or bases
added dropwise along the side of the vessel, to
avoid any denaturation in adjustment of pH.

The lifetime of many enzymes can be greatly pro-
longed by cold storage. For most enzymes a storage
in a refrigerator at $2-5^{\circ}C$ is sufficient for long
term stability in the dry state. Other enzymes are
unstable even at $2-5^{\circ}C$, and must be stored in a
freezer well below $0^{\circ}C$. Some enzymes are stabilized
at high concentrations of salts and can be kept for
long periods as suspensions in ammonium sulfate.
Such solutions can often be stored in a refrigerator
or freezer for months without loss of activity, although repeated freezing and thawing is to be avoided.

Organic solvents, i.e. alcohol, acetone, ether,
denature most enzymes at room temperature, except at
low concentrations (less than 3%). Care must be
taken in changing the composition of a solution from
aqueous to partly nonaqueous.

In some cases care must be taken to avoid the presence of air or oxygen in storage of enzyme solutions. The sulfhydryl group in CoA is easily oxidized by atmospheric oxygen, for example, while NADH
and NADPH must be protected from light and stored in
a desiccator in the cold. NADPH has about 10 times
greater stability when stored in the dark than when
stored in direct sunlight.

Many enzymes are denatured at surfaces; therefore,
the formation of froth should be avoided. The vigo-
rous shaking of an enzyme solution in dissolution
is poor practice. Likewise, stirrers which whip up
the surface should be avoided.

Some buffer solutions, especially phosphate, are
good growth media for micro organisms. These should

be carefully handled and stored. To avoid contamina-
tion, each day's supply of buffer should be removed
by pouring and not by pipetting.

Because traces of metal ions can cause a loss of
enzymic activity, it is necessary that the water
used in preparation of reagents, etc., be carefully
purified, either by deionization or by a double
distillation from $KMnO_4$ in an all glass still. All
pipettes, flasks and containers should be thoroughly
cleaned. After a double rinse with distilled water,
pipettes should be dried in an oven at a moderate
temperature (never by rinsing with acetone or other
organic solvents).

F. SOURCES OF REAGENTS

One of the often cited objections to enzymic
methods has been the instability and unavailability
of enzymes. However, today there are many companies
that sell good enzyme products. A listing of manu-
facturers and the enzyme products sold can be found
in the Appendix. Substrates are available from
Aldrich (Milwaukee), Eastman (Rochester, N. Y.),
Fisher Scientific (Pittsburgh), Pierce Chemical
(Rockford, Illinois), Calbiochem, Sigma, Miles, NBC,
Mann and Worthington

G. ENZYME ACTIVITY

Unlike conventional inorganic or organic analytical reagents which vary between 90-100% purity,
the purity of most enzymes generally varies between
1 and 5%. It is therefore essential to define an
arbitrary unit of enzyme, in terms of which the
purity and activity can be expressed quantitatively.
In most cases this unit is defined with reference
to the method used. In a spectrophotometric method,
it is the amount of enzyme which produces a certain

change of a particular substrate (usually the
natural substrate for an enzyme, i.e. uric acid for
uricase, etc.) per minute. Unfortunately, many
different units have been proposed for the same enzyme
by different methods. Attempts have been made, however, to standardize the unit for a particular enzyme
in order to avoid confusion. In 1959 the Enzyme
Commission of the International Union of Biochemistry
and the Clinical Chemistry Commission of the International Union of Pure and Applied Chemistry accepted
the proposal of Racker et al.[23] for adoption of an
"International Unit." This unit is defined as the
amount of enzyme which converts 1 μ mole of substrate
per minute at $25^{o}C$, optimum substrate concentration,
optimum ionic strength of buffer and optimum pH.
The specific activity of the enzyme is defined as
the activity of one milligram of the enzyme.

The "International Unit" represents an important
advance in enzymology and will allow an easy comparison of different enzyme preparations or the same
preparation from different sources.

REFERENCES

1. E. H. White, F. McCapra, G. F. Field, J. Am.
    Chem. Soc. 85, 337 (1963).

2. S. P. Colowick and N. O. Kaplan, eds., Methods
    in Enzymology, Academic Press, New York, 1957.

3. G. Osann, Poggendorfs Ann. 67, 372 (1845).

4. O. Warburg, Wassenstoffubertragende Fermente,
    Varlag, Berlin, 1948.

5. H. Lineweaver and D. J. Burk, J. Am. Chem. Soc.
    56, 658 (1934).

6. E. Schütz, Z. Physiol. Chem. 9, 577 (1885).

7. J. Schütz, Z. Physiol. Chem. 30, 1 (1900).

8. B. S. Millar and J. A. Johnson, Arch. Biochem.
    Biophys. 32, 200 (1951).

9. A. B. Roy, Biochem. J. 57, 465 (1954).

10. P. Baum and R. Czok, Biochem. Z. 332, 121 (1959).

11. E. Adler, G. Gunther and M. Plass, Biochem. J. 33, 1028 (1939).

12. W. J. Blaedel and G. P. Hicks in Advances in Analytical Chemistry and Instrumentation, Vol. 3 (C. N. Reilley, ed.), Interscience, New York, 1964, pp. 105-140.

13. G. G. Guilbault, D. N. Kramer and P. L. Cannon, Anal. Chem. 34, 1437 (1962).

14. H. Linde, Anal. Chem. 31, 2092 (1959).

15. C. McGaughey and E. Stowell, Anal. Chem. 36, 2344 (1964).

16. K. B. Yatsimirskii, Kinetic Methods of Analysis, Pergamon, Oxford, 1966.

17. G. G. Guilbault, Kinetic Methods of Analysis, in Fluorescence. Theory, Instrumentation and Practice, (G. G. Guilbault, ed.), Marcel Dekker, Inc., New York, 1967, pp. 306-307.

18. B. F. Erlanger, S. N. Burbaum, R. A. Sack and A. G. Cooper, Anal. Biochem. 19, 542 (1967).

19. G. G. Guilbault, Kinetic Methods of Analysis, in Fluorescence. Theory, Instrumentation and Practice, (G. G. Guilbault, ed.), Marcel Dekker, Inc., New York, 1967, pp. 297-358.

20. G. G. Guilbault, P. Brignac and M. Zimmer, Anal. Chem. 40, 190 (1968).

21. G. G. Guilbault, P. Brignac and M. Juneau, Anal. Chem. 40, 1256 (1968).

22. H. Pardue, M. Burke and D. O. Jones, J. Chem. Ed. 44 (11), 684 (1967).

23. J. Cooper, P. A. Srere, M. Tabachnick and E. Racker, Arch. Biochem. Biophys. 74, 306 (1958).

CHAPTER 2

METHODS OF ASSAY

A.   CHEMICAL METHODS

To follow the progress of an enzymic reaction, one
must monitor the change with time of the concentration of either one of the reactants or one of the products of the reaction.

In a chemical method, .one follows the reaction by
periodically drawing out a sample from the reaction mixture and measuring one of the reactants or products by a
volumetric procedure.   For example, in the enzymic
reaction:

$$\text{Hydrogen peroxide} \xrightarrow{\text{catalase}} H_2O + O_2$$

the extent of reaction could be determined by periodically titrating a sample of hydrogen peroxide with
cerium(IV).   Generally one must stop the reaction in
the sample drawn in order to analyze for a reactant
or product, especially if the reaction is proceeding
at an appreciable rate.   One does this by 1) adding
a substance which either combines with one of the
reactants or inhibits the reaction, 2) cooling the
reaction, or 3) if the reaction is pH dependent, by
suddenly adding acid or alkali to change the pH.

B.   INSTRUMENTAL METHODS

Generally it is more desirable to be able to continuously monitor a chemical reaction without having
to draw and titrate samples.   One can do this by 1)
following the appearance or disappearance of some

LIBRARY
College of St. Francis
JOLIET, ILL.

73704

species by monitoring its physiochemical properties
directly or 2) using a coupled reaction sequence.
   One might monitor the reaction:

      Substrate + Enzyme ──────$>$ Products

by following the change in absorbance of the system,
if any reactant or product is colored. Or the reaction can be monitored electrochemically if either a
reactant or product is electroactive. The change in
pH could be recorded and equated to enzymic activity,
if H^+ is one of the reactants or products.

1. Manometric Methods

   If one of the products of an enzyme reaction is a
gas, the extent of reaction could be indicated by a
volumetric measurement of the gas produced using a
manometer. Such techniques, called manometric
methods, were originally proposed by Bancroft[1] and
were developed by Warburg.[2]   Such techniques have
been extensively used in the measurement of: 1) gas
consuming reactions in which the O_2 uptake is measured
(e.g. oxidative enzyme systems, such as glucose oxidase,
peroxidase, cytochrome oxidase, etc.); 2) gas producing
reactions, in which one of the products of the enzymic
reaction is a gas (e.g. CO_2 found in a decarboxylase
enzyme system, or NH_3 formed from urea in the urease
system); or 3) acid forming enzymic reactions which
are carried out in the presence of bicarbonate in
equilibrium with a gas mixture containing a definite
percentage of CO_2. In this technique any acid produced in the enzyme reaction will react with the bicarbonate to give a corresponding amount of CO_2, which
can be measured on the manometer. Since all NAD
(nicotinamide adenine dinucleotide) dependent dehydrogenase reactions yield a proton in the reduction
of NAD, manometric methods have been extended to
these enzyme systems also. A detailed discussion on

manometric techniques can be found in books by Umbreit et al.[3] and by Dixon.[4]

2. Spectrophotometric Methods

If either one of the reactants or one of the products
of an enzyme reaction is absorbing either in the ultra-
violet, visible or infrared region of the spectra,
then it is possible to monitor the progress of such an
enzyme reaction spectrophotometrically. Consider for
example a typical reaction A $\xrightarrow{\text{Enzyme}}$ B + C. The UV-
visible spectra of A, B and C shown in Fig. 1 indi-
cates that A has a strong absorbance in the UV, with a
λ_{max} of 295 mμ. Products B and C absorb strongly in
the visible with λ_{max} of 420 and 560 mμ, respectively.
One might follow the progress of this enzymic reaction
by noting the decrease in absorbance of A at 290 mμ, or
by the increase in absorbance at 420 or 560 mμ as B and
C are formed. Experimentally the reaction can best be
monitored at 560 mμ since both B and C absorb at 420
mμ and generally it is better to follow an increase
in absorbance than a decrease. Furthermore, C has a
higher molar absorbtivity (absorbance per mole per
unit cm path length) than A, so a greater sensitivity
can be realized. Today there are many good spectro-
photometers available that cover the entire UV-visible-
near IR region of the spectra. Generally an instrument that reads the change in absorbance with time
automatically like the Beckman DB or Cary 14 is preferred to a null point instrument like the Beckman DU
which requires a point by point plot of absorbance
changes. Likewise a double beam instrument is better
than a single beam, since all measurements can be
made against a cuvette containing all the reagents of
an assay mixture except one (reagent blank).

In studying enzyme reactions spectrophotometrically,
it is essential that the cell containing the reacting
mixture be thermostatically controlled, since a change

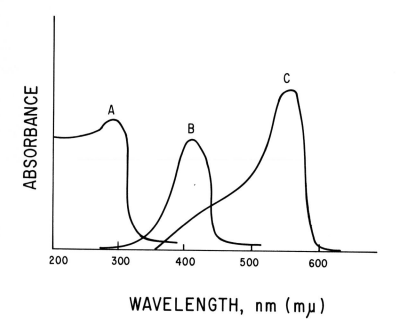

WAVELENGTH, nm (mμ)

FIG. 1

in temperature of $1^{\circ}C$ will cause approximately a 10%
change in the rate. The temperature of the cell must
be controlled within $\pm$ $0.2^{\circ}C$. Since most instruments
are not built with such a thermostat, it is necessary
to have a jacketed cell-holder through which water can
be circulated from an external thermostated water bath.

It was the evolution of spectrophotometric methods
three decades ago that proved a boon to enzymatic
methods; starting when Warburg showed that reduced
coenzymes, NADH and NADPH, absorb at 340 mμ. The oxidized coenzymes absorb at 270 mμ but not at 340 mμ,
thus providing a method for the assay of dehydrogenase
systems. A typical curve for the measurement of the
activity of a dehydrogenase is indicated in Fig. 2.
Following the change in absorbance at 340 mμ due to
production of NADH, the activity of a dehydrogenase,
such as lactic dehydrogenase in serum, can be calculated. Other examples of enzymic reactions that can
be followed spectrophotometrically are the xanthine
oxidase conversion of a non-absorbing hypoxanthine to
a highly absorbing, conjugated ring compound, uric
acid, λ_{max}=290 mμ; the assay of uric acid with uricase, following the decrease in absorbance at 290 mμ;
and the assay of cholinesterase by monitoring the
yellow color of a thiocholine bisdithionitrobenzoate
complex:

$$\text{Acetylthiocholine} \xrightarrow[\text{Bisdithionitrobenzoic acid}]{\text{ChE}}$$

$$\text{Yellow Complex}$$

In many cases an appreciable absorption change is
not observed in the enzyme reaction being studied.
In such a case a coupled reaction sequence is used,
with a second reaction used to "indicate" the progress
of the enzyme reaction. For example, glucose is
catalytically oxidized in the presence of glucose oxidase to peroxide. None of the reactants or products

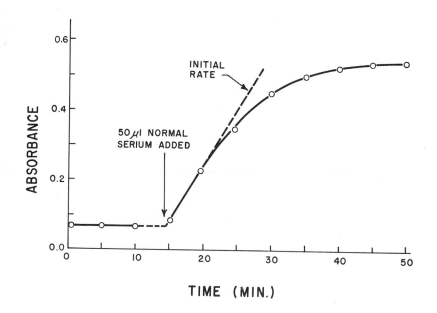

FIG. 2

Analysis of lactic dehydrogenase in serum
using lactate dehydrogenase. Absorbance
change at 340 mμ is measured.

are absorbing. Yet glucose can be easily assayed by using an indicator reaction and following the rate of formation of a colored dye:

Enzyme reaction:

$$\text{Glucose} + H_2O_2 + O_2 \xrightarrow{\text{Glucose Oxidase}} \text{Gluconic Acid} + H_2O_2$$

Indicator reaction:

$$H_2O_2 + \text{leuco-dye} \xrightarrow{\text{Peroxidase}} H_2O + \text{dye}$$

3. Polarimetric Methods

Since many enzymes catalyze the reaction of only one optical isomer of a substrate, yielding an optically inactive product, the reaction can be followed by the change of optical rotation. Likewise, the production of an optically active isomer from an optically inactive substrate can be easily monitored. The activity of sugar enzymes, such as sucrase, can be nicely monitored by such a technique which requires only a commercially available polarimeter with a thermostated tube.

In those cases where the substrate or product has too low a rotation to measure directly, the rotation may be increased by forming a complex. Lactic acid, for example, and other hydroxy acids form strong complexes with molybdate that have high specific rotations. Many dehydrogenases can be monitored by these techniques.

4. Electrochemical Methods

a. Ion Selective Electrodes. Probably the most common electrochemical method that has been used in enzymology is one using a glass electrode in following reactions which involve the production of acid. Because changes in pH affect the activity of the enzyme and also the rate of reaction, direct readings of pH changes are generally not used. Instead a "pH stat"

method is generally employed, in which the pH is
maintained at a constant value by frequent addition
of alkali. The rate at which base is added then gives
the reaction velocity independent of the amount of
buffer.

Several convenient automatic "pH stat" instruments are available, probably the most common being
the Radiometer (Stockholm) and the Metrohm (Brinkman
Instrument Co.). Both these instruments maintain a
constant pH by continuous automatic additions of acid
or alkali, and at the same time automatically record
the amount added as a function of time.

The oxygen electrode has found increasing use in
the enzymic analysis of oxygen consuming enzymic systems:

$$\text{Substrate} + O_2 \xrightarrow{\text{Enzyme}} \text{Oxidized substrate}$$

The electrode consists of a gold cathode separated
by an epoxy casting from a tubular silver anode.
The inner sensor body is housed in a plastic casing
and comes in contact with the solution only through
the membrane. When oxygen diffuses through the membrane it is electrically reduced at the cathode by an
applied potential of 0.8 volts. This reaction causes
a current to flow between the anode and cathode which
is proportional to the partial pressure of oxygen in
the sample. Oxygen electrodes are available commercially or can be easily prepared in the laboratory.[5]
Of the various techniques available for monitoring
the glucose content of blood, many researchers feel
the oxygen electrode method is the most reliable.
Kadish and Hall[6], Makino and Koono[7] and Sternberg et al.[8] found a good correlation between glucose values determined in blood by a measurement
of oxygen uptake with those obtained by standard
chemical tests.

b. _Potentiometry at Small Current_. Potentiometric, ampeometric and polarographic techniques
have been wisely used by analysts to follow enzyme
activity. Guilbault et al.[9] have proposed a kinetic
method for enzyme reactions based on the electrochemical measurement of the rate of cleavage of a substrate by the enzyme sample. Rates were measured by
recording the difference in potential between two
platinum electrodes polarized with a small, constant
current. Any reaction of the type $A \xrightarrow{B} C + D$, where
the substrate A undergoes enzymolysis by B to form
products C and D, can be followed provided C and/or
D are either more or less electroactive than substrate A. For example, in the cholinesterase catalyzed hydrolysis of thiocholine esters, a thiol is
produced upon enzymic hydrolysis which is more electroactive (has a lower oxidation potential) than the
substrate. A reduction in potential results (Fig. 3),
and plots of ΔE/min. vs. cholinesterase activity yield
straight line calibration plots:

$$R-C-O-S-R' + H_2O \xrightarrow{\text{cholinesterase}} R'-SH + R-COOH$$

Pesticides, such as Sarin, Systox, Parathion and Malathion, which inhibit cholinesterase, can be determined
at 10^{-9}g concentrations by this technique, with deviation of 1%.[10] Other enzyme systems such as glucose-
glucose oxidase[11] xanthine oxidase[12] and peroxidase and catalase[13] can be determined by the electrochemical technique. Care must be taken to rinse
the electrodes thoroughly after each use, since proteins adsorb on Pt and will cause a decrease in sensitivity after a number of measurements.

c. _Amperometry_. In amperometric methods, a constant potential is applied between 2 electrodes
immersed in a solution of the material to be analyzed.

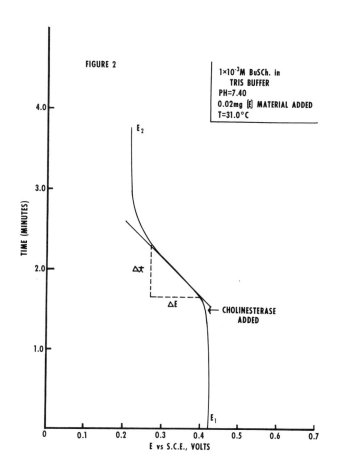

FIG. 3

Voltage - time curves for enzymatic hydrolysis
of butyrylthiocholine iodide by cholinesterase.
(ref. 9)

The change in the current is then recorded with
change in reaction conditions (time, addition of
reagent, etc.). For example, in the assay of glucose with glucose oxidase, Blaedel and Olson[14]
measured the change in current that resulted upon
oxidation of ferrocyanide to ferricyanide at a tubular platinum electrode:

$$\text{Glucose + Glucose Oxidase} \longrightarrow \text{Peroxide}$$
$$\text{Peroxide + Ferrocyanide} \longrightarrow \text{Ferricyanide}$$

The total current which is proportional to the
relative amount of ferricyanide and ferrocyanide
present is measured and equated to the concentration
of glucose present. Pardue[15,16] utilized a similar
system for the analysis of glucose and galactose,
except that iodide was used instead of ferrocyanide.
Again the total current is proportional to the relative amounts of iodide and iodine, and hence to the
amount of galactose or glucose present.

$$\text{Galactose + Galactose Oxidase} \longrightarrow \text{Peroxide}$$
$$\text{Peroxide + Iodide} \xrightarrow{\text{Mo(VI)}} \text{Iodine}$$

d. <u>Coulometry</u>. Coulometry has also found considerable use in enzymic methods of assay. Coulometric methods are based on the exact measurement of
the quantity of electricity that passes through a
solution during the occurence of an electrochemical
reaction. The component to be determined may be
either oxidized or reduced at one of the electrodes
(primary coulometric analysis) or may react quantitatively in solution with a single product of electrolysis (secondary coulometric analysis).

Purdy, Christian and Knoblock,[17] for example, described a method for the analysis of urea based on the
urease hydrolysis of urea to form ammonia. The result-

ing ammonia is then titrated with coulometrically
generated hypobromite using a direct amperometric end
point. Simon, Christian and Purdy[18] described a
coulometric method for glucose in human serum. Glucose
oxidase specifically catalyzes the aerobic oxidation
of glucose to hydrogen peroxide; the peroxide reacts
with iodide in the presence of Mo(VI) catalyst to
form iodine. A known excess of thiosulfate reduces the
iodine as it is produced and the excess thiosulfate is
titrated coulometrically with electrogenerated iodine.

e. Polarography. Polarographic methods (in which
the change in diffusion current is recorded with change
in the potential applied) have found considerable use
in enzymic analysis. Cholinesterase has been determined by a measurement of the change in current resulting from the production of thiol from acetyl thiocholine iodide effected by cholinesterase.[19] Catalase[20,21] and 3-hydroxy anthranilic oxidase[22] have
been determined by similar polarographic techniques.

A thorough discussion of the advantages and disadvantages of polarography in biochemical analysis
can be found in a book by Purdy.[23] Also discussed
are interferences and problems associated with polarography and other electrochemical methods.

5. Fluorescence Methods

Because of limitations in molar absorptivities,
measurements of gas volumes, or of changes in pH,
most methods previously described for measuring
components in enzyme reactions are limited to reactions of reagents present at concentrations greater
than 10^{-6}M. Because fluorometric methods are generally
several orders of magnitude more sensitive than chromogenic ones, a large increase in the sensitivity of
measurement should result. Thus, much lower concentrations of reactants would be needed and one could

devise methods for substances at $10^{-9}M$ concentrations and lower. Moreover, fluorometric methods are quite useful in biochemical work in the localization of enzymes, related substrates and coenzymes, within organs and even within individual cells.

A schematic diagram of a fluorometer is indicated in Fig. 4. Light from a suitable source passes through a filter or a monochromater (primary or excitation filter or monochromator) and impinges on the sample. That portion of light that is emitted (measured at right angles to eliminate measurement of any transmitted light) is passed through a secondary filter or monochromator and onto the photodetector. The signal from the detector is amplified and can be measured on a meter or recorder.

Fluorescence measurements are generally several orders of magnitude more sensitive than colorimetric ones since in fluorescence one measures an increase in signal over a zero background, while in spectrophotometry, a decrease in a large standing current is measured. This fluorescence signal is a maximum when the optimum wavelengths for excitation and emission are used.

The instrumentation for fluorescence is very similar to that used in spectrophotometry, differing only in a right angle measurement rather than a straight line one, and in the use of a second filter, or monochromator. In fact almost any commercial spectrophotometer like the Beckman DU or DK can be easily converted to a fluorometer. In addition many companies today sell both filter and monochromator (grating or prism) fluorometers: Turner, American Instrument, Farrand, Zeiss, Baird Atomic, etc.

Because of their sensitivity and specificity fluorescence methods have found increasing usage in enzymology. For example, the reduced forms of

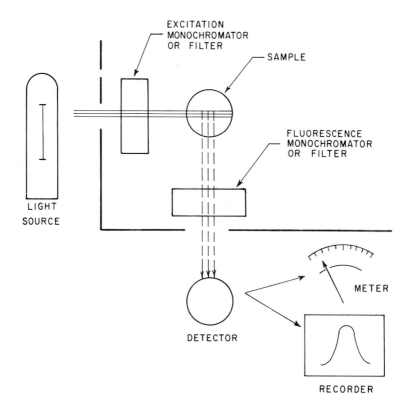

FIG. 4
Schematic of a Typical Fluorometer.

nicotinamide adenine dinucleotide, NADH, and nicotinamide adenine dinucleotide phosphate, NADPH, are highly
fluorescent. Thus all NAD and NADP-dependent reactions
involved in enzymatic analysis can be measured fluorometrically, with an increase of 2-3 orders of magnitude
in sensitivity over colorimetric techniques。 Fluorescence methods have also been used extensively for the
determination of hydrolytic enzymes, based on the enzyme catalyzed hydrolysis of a non-fluorescent ester
to a highly fluorescent alcohol or amine. Guilbault
and Kramer[24,25], for example, described a rapid,
simple method for the determination of lipase, based
on its hydrolysis of the non-fluorescent dibutyryl
ester of fluorescein.

   Fluorescein is produced upon enzymolysis, which is
highly fluorescent.

$$\text{Dibutyryl Fluorescein} \xrightarrow{\text{Lipase}} \text{Fluorescein}$$
$$\text{(Non-fluorescent)} \qquad\qquad \text{(Fluorescent)}$$

This reaction can be monitored by measurement of the
rate of production of the highly fluorescent fluorescein with time, $\Delta F/min$. The concentration of enzyme
can then be calculated from linear calibration plots of
$\Delta F/min$ vs. enzyme concentration.

   Fluorescent quenching is one of the most serious
problems associated with the use of fluorescence in
enzymic analysis. Highly absorbing molecules (i.e.
dichromate) rob energy from the molecule under study,
lowering the total fluorescence observed. Other interferences are molecules that absorb or fluorescence at the
same wavelengths as the substance being determined.
Proteins, for example, are serious interferences in
fluorescence measurements made in the ultra violet
region because they contain amino acids (i.e. tryptophan and tyrosine) that are fluorescent in this region.
For this reason it is better to make fluorescence

measurements in the visible region whenever possible.
Hence a fluorogenic substrate that is cleaved to a red
fluorescent compound would be preferred to one that
yields a blue fluorescent substance.[26,27]

6. Radiochemical Methods

The activity of an enzyme can be measured using a
radio-actively "tagged" substrate, which upon enzymolysis yields a radioactive product. The amount of
radioactive product formed with time, is then proportional to the concentration of enzyme.

Acetyl-1-C^{14} choline, for example, has been employed
as a substrate for acetylcholinesterase by Reed, Goto
and Wang[28] and Potter.[29] After a removal of unhydrolyzed substrate by ion exchange, the acetic acid-1-C^{14}
formed by enzymic activity is measured.

$$CH_3-\overset{+}{\underset{\underset{CH_3}{/|}}{N}}-CH_2-CH_2-O-\underset{\underset{O}{\|}}{C}-C^{14}H_3 \longrightarrow CH_3-\overset{+}{\underset{\underset{CH_3}{/|}}{N}}-CH_2-CH_2-OH + HO\underset{\underset{O}{\|}}{C}-C^{14}H_3$$

Many radioactive substrates are available commercially from companies such as New England Nuclear.
Radioactivity can be measured with an instrument as
simple as the Vanguard 4 π paper strip counter. Since
both the substrate and product are radioactive, a prior
separation of the two must be effected before measurement. This can usually be done by distillation or
chromatography.

REFERENCES

1. J. Bancroft, J. Physiol. <u>37</u>, 12 (1908).

2. O. Warburg, Biochem. Z. <u>152</u>, 51 (1924).

3. W. W. Umbreit, R. H. Burris and J. F. Stauffer,
   Manometric Techniques, Burgess Publishing Co.,
   Minneapolis, 1945.

4. M. Dixon, Manometric Methods, Cambridge University
   Press, England (1951).

5. G. Charlton, D. Read and J. Reed, J. Appl. Physiol. 18, 1247 (1963).

6. A. H. Kadish and D. A. Hall, Clin. Chem. 9, 869 (1965).

7. Y. Makino and K. Koono, Rinsho Byori 15, 391 (1967) (Japan).

8. A. Kadish, R. Litle, J. C. Sternberg, Clin. Chem. 14, 116 (1968).

9. G. G. Guilbault, D. N. Kramer and P. L Cannon, Anal. Chem. 34, 842 (1962).

10. Ibid, p. 1437.

11. G. G. Guilbault, B. Tyson, D. N. Kramer and P. L. Cannon, Anal. Chem. 35, 582 (1963).

12. G. G. Guilbault, D. N. Kramer and P. L. Cannon, Anal. Chem. 36, 606 (1964).

13. G. G. Guilbault, Anal. Biochem. 14, 61 (1966).

14. W. J. Blaedel and C. Olson, Anal. Chem. 36, 343 (1964).

15. H. Pardue, Anal. Chem. 35, 1240 (1963).

16. H. Pardue, R. Simon, Anal. Biochem 9, 204 (1964).

17. W. C. Purdy, G. D. Christian, E. C. Knoblock, Presented at the Northeast Section, American Association of Clinical Chemists, 16th National Meeting, Boston, Mass., August 17-20, 1964.

18. R. K. Simon, G. D. Christian and W. C. Purdy, Clin. Chem. 14, 463, (1968).

19. V. Fischerova-Bergerova, Pracovni Lekarstvi 16, (1) 8 (1964).

20. T. H. Ridgway and H. B. Mark, Anal. Biochem. 12, 357 (1965).

21. H. Jacob, Z. Chem. 4, 189 (1964).

22. M. N. Gadaleta, E. Lofrumento, C. Landriscina, A. Alifano, Bull. Soc. Ital. Biol. Sper. 39, (24) 1866 (1963).

23. W. C. Purdy, Electroanalytical Methods in Biochemistry, McGraw Hill, New York, 1965.

24. G. G. Guilbault and D. N. Kramer, Anal. Chem. 35, 588 (1963).

25. Ibid, ,36, 409 (1964).

26. G. Guilbault, Fluorescence. Theory, Instrumentation and Practice, Marcel Dekker, Inc., New York, 1967.

27. S. Udenfriend, Fluorescence Assay in Biology and Medicine, Academic Press, New York, 1963.

28. D. L. Reed, K. Goto and C. H. Wang, Anal. Biochem. 16, 59 (1966).

29. L. T. Potter, J. Pharm. Exp. Ther. 156, 500 (1967).

CHAPTER 3

DETERMINATION OF ENZYMES

A. GENERAL

The determination of the activity of an enzyme has a wide range of applications. Enzyme assays are of importance in the field of food, agricultural, forensic and clinical chemistry, especially in the detection of various diseases of the body. Good enzyme assay procedures are necessary to form the basis for acceptable analytical techniques for the analysis of substrates, activators and inhibitors. The poor precision, slowness and labor that have made enzyme catalyzed reactions unappealing as a means of analysis have been more a consequence of poor procedures than the fault of the enzymes. With the advent of new techniques, fluorometric and electrochemical, many of the previous difficulties have been resolved.

In this chapter some of the methods available for the assay of those enzymes of most interest to analytical biochemists and clinical chemists will be discussed.

B. HYDROLYTIC ENZYMES

1. Amylase

Probably the easiest and most accurate method for the measurement of amylase activity is one involving a determination of the reducing sugar liberated by the enzyme. In the procedure of Bernfeld[1] the

maltose liberated from starch is measured by its
ability to reduce 3,5-dinitrosalicylic acid. A unit
of α- or β- amylase activity is that liberating a
micromole of α- or β-maltose per minute at $25^{o}C$ from
1% soluble starch in 0.016M acetate buffer, pH 4.8.
The color produced from a chromogenic reagent (an
alkaline solution of 3,5-dinitrosalicylic acid and
potassium-sodium tartrate) at 540 mμ is proportional
to the amylase activity. This method can give inaccurate results with serum when the blood sugar level is
above 150 mg.[2]

Other methods for measuring amylase activity include a determination of residual substrate by a
precipitation with ethanol-water, and a colorimetric
measurement of the changes in the iodine color of the
assay mixture.[3] In the latter method the substrate
is incubated with amylase for 15 minutes. The amount
of decrease in the absorbance of iodine at 620 mμ, due
to a decolorization of the iodine by reducing sugar,
is a measure of the activity of amylase present.

2. Cholinesterase

Cholinesterases (ChE) catalyze the hydrolysis of
choline esters:

$$\text{Acylcholine} + H_2O \xrightarrow{\text{ChE}} \text{Choline} + \text{acid}$$

The best known enzymes in the cholinesterase family
are acetylcholinesterase (AChE) and butyrylcholin-
esterase (BuChE). These enzymes differ on the basis
of substrate selectivity: both catalyze the hydrolysis
of acetylcholine(I), but the rate with AChE is faster.
AChE is inactive on butyrylcholine, however, whereas
BuChE rapidly hydrolyzes this ester. A common name
for AChE is "true" cholinesterase.

a. Manometric and pH Methods. Perhaps more analytical methods have been proposed for cholinesterase than for any other enzyme. Classical methods
for this enzyme involved a measurement of the acid

produced during hydrolysis of acetylcholine(I), either
manometrically by liberation of CO_2 or by potentiometric titration. In either method one unit is equal
to one micromole of acetylcholine hydrolyzed per
minute. In the manometric method the amount of acetylcholine hydrolyzed is determined directly using the
Warburg apparatus.[4] The acid liberated releases an
equivalent amount of CO_2 from a bicarbonate-CO_2
buffer. In the potentiometric method a pH stat or an
ordinary pH meter with external electrodes can be used
to monitor the acid produced. In the pH stat approach
the amount of base required to maintain a constant
pH is recorded, and the rate of addition of base is a
measure of the rate of enzymic hydrolysis. A typical
curve obtained for the hydrolysis of acetylcholine
using the pH stat is indicated in Fig. 1.

　　b. <u>Colorimetric Methods</u>. Cholinesterase can be
measured colorimetrically using the indophenyl acetate
substrate proposed by Kramer and Gamson.[5] The rate
of production of the blue indophenol is a measure of

Indophenyl acetate $\xrightarrow{\text{ChE}}$ Indophenol + acetic acid
　　　(Red)　　　　　　　　　　(Blue)

the activity of cholinesterase. This substrate as well
as other colored substrates have been used on paper
strips for the semiquantitative assay of serum cholinesterase.[5]

　　Another colorimetric method for cholinesterase involves the formation of a brown ferric hydroxamate
complex(II) with unhydrolyzed acetylcholine(I) substrate.[6]

$(CH_3)_3\overset{\oplus}{N}$ $CH_2-CH_2-O\underset{\underset{O}{\|}}{C}-CH_3$ + NH_2OH ——————$>$

　　　I　　　　　O

　　　　　　$(CH_3)_3\overset{\oplus}{N}$ CH_2CH_2-OH + $CH_3-\underset{\underset{O}{\|}}{C}-NHOH$

　　　　　　　　　　　　　　　　　　　　$\downarrow Fe^{3+}$

　　　　　　　　Complex(λ_{max}=540 mμ)
　　　　　　　　　　II

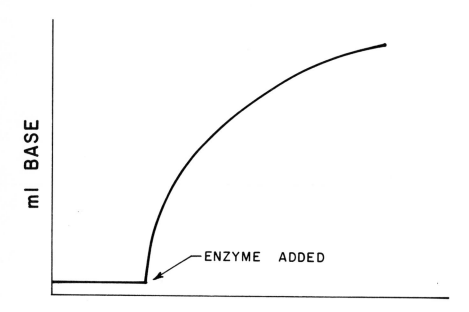

FIG. 1

Rate Curves for the Splitting of Acetylcholine
by Cholinesterase measured with Δ pH Control.

Ellman[7] has developed a simple colorimetric
procedure of assay of cholinesterase that involves
a reaction of thiol, produced in the enzymic hydrolysis of the thiol analog of acetylcholine(III),
with bisdithio nitrobenzoate (BDTNB). A deep yellow
color is formed which can be measured at 340 mμ.

$$(CH_3)_3\overset{+}{N}(CH_2)_2S\text{-}\overset{O}{\underset{\|}{C}}\text{-}CH_3 \xrightarrow{ChE} (CH_3)_3\overset{+}{N}(CH_2)_2\text{-}S\text{-}H$$

III   BDTNB
Yellow Complex (λ_{max}=340 mμ)

   c. <u>Electrochemical Methods</u>. Guilbault, Kramer
and Cannon[8] proposed an electrochemical method for
the determination of cholinesterase, based on the
hydrolysis of acetyl thiocholine iodide(III) by cholinesterase. A small, constant current of 25 μa
(Fig. 2) is applied across two platinum thimble electrodes, and the change in potential with time upon
hydrolysis is recorded. A typical curve, illustrated
in Fig. 3, Chapter 2, results. Initially, a constant
potential is obtained, due to the oxidation of iodide
of the substrate to iodine.[9] When cholinesterase is
added, the potential drops, due to the formation of the
more electrochemically active thiol(IV). The potential changes are measured with a high impedance vacuum
tube voltmeter (VTVM).

$$(CH_3)_3\overset{+}{N}\text{-}(CH_2)_2\text{-}S\text{-}\overset{O}{\underset{\|}{C}}\text{-}CH_3 \underset{I^-}{} \xrightarrow{ChE} (CH_3)_3\overset{+}{N}\text{-}(CH_2)_2\text{-}SH \underset{I^-}{}$$

III   IV
$E_o \sim 0.45v$ $E_o \sim 0.2$ v

The slopes of the resulting depolarization curves
provide data on the rates of enzymatic hydrolysis of
the thiocholine esters. These rates correspond well
with those predicted by Michaelis-Menten kinetics. By
this procedure 0.2 to 14 units of cholinesterase
could be assayed with a standard deviation of 0.7%.
The complete theory of this method has been worked
out.[9,10]

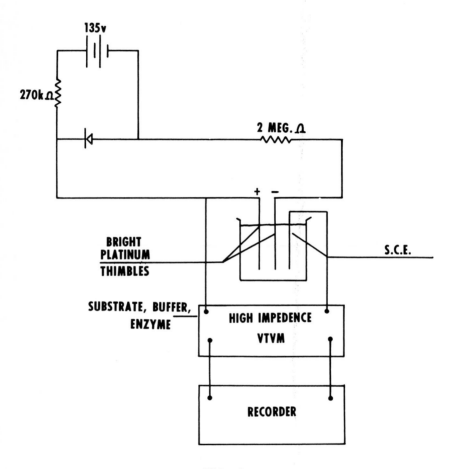

FIG. 2

Constant Current Apparatus
(ref. 8)

d.   Fluorescence Methods. Because of limitations
in molar absorptivities, measurements of gas volume or
of changes in pH, most procedures for cholinesterase
are limited to substrate concentrations greater than
10^{-5}M and enzyme concentrations greater than 10^{-2}
units. Since fluorogenic substrates are generally
several orders of magnitude more sensitive than chromogenic ones, a large increase in the sensitivity of
enzymic assay should result from the use of esters of
fluorogenic materials. These compounds, themselves
non-fluorescent, could be hydrolyzed by enzymes to
form easily measured fluorescent products. The production of the fluorescence can be followed kinetically
and equated to enzyme activity.

Guilbault and Kramer prepared four fluorogenic substrates for cholinesterase: resorufin butyrate[11],
indoxyl acetate[12] and α- and β-naphthyl acetate[13].
Resorufin esters are non-fluorescent, but are hydrolyzed to resorufin(V) which is highly fluorescent.
Indoxylacetate is non-fluorescent, but is hydrolyzed
by cholinesterase first to indoxyl(VI), then indigo
white(VII), both of which are highly fluorescent
(λ_{ex} = 395 mμ and λ_{em} = 470 mμ). At pH values less
than 7, the highly fluorescent indigo white is formed
which is stable with time and is not air oxidized to
indigo blue (VIII). At more alkaline pH values the
fluorescence of the solution rapidly decreases with
time. Both indoxyl and indigo white have the same
fluorescence excitation and emission wavelengths.
The fluorescence intensity of indigo white is twice
that of indoxyl.

With either resorufin butyrate or indoxyl acetate
as substrate, from 0.0003 to 0.12 units per ml of
cholinesterase could be assayed in 2-3 minutes with
an accuracy and precision of about 1%.

c

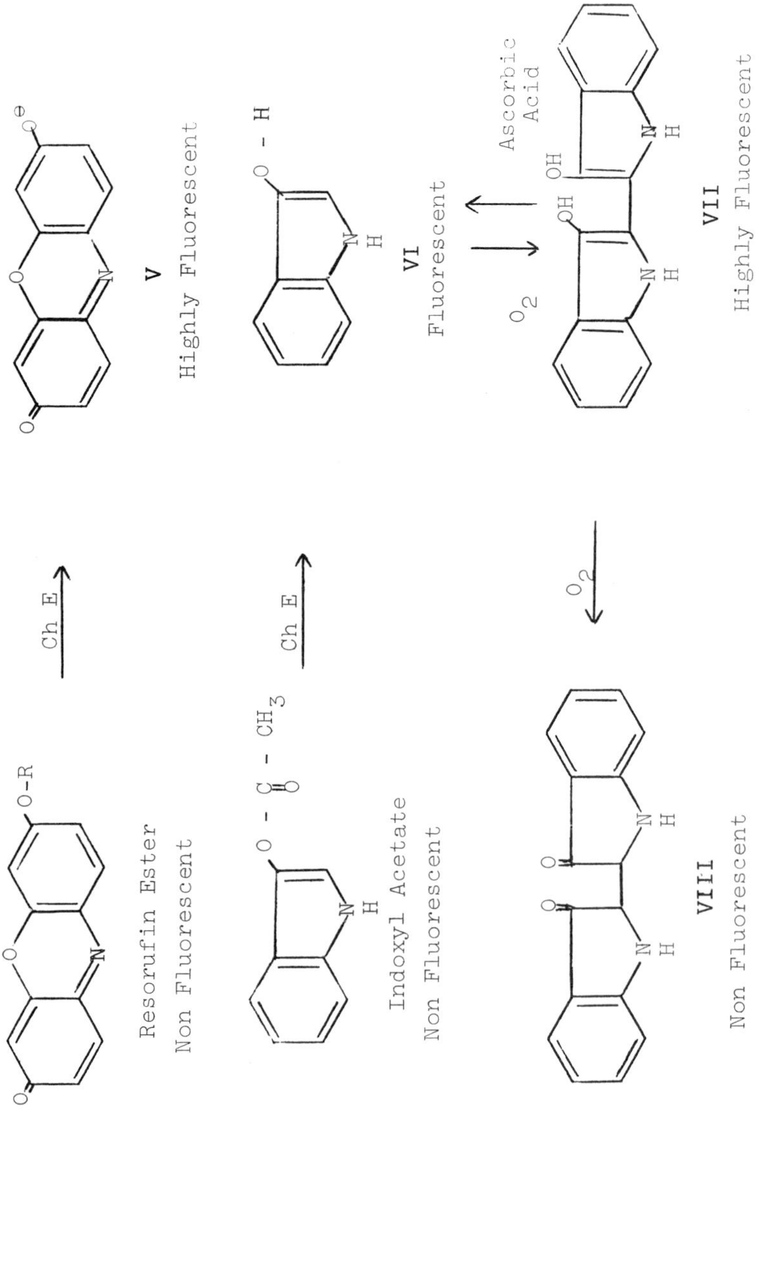

V

Highly Fluorescent

VI

Fluorescent

Ascorbic Acid

O_2

VII

Highly Fluorescent

Ch E

Resorufin Ester

Non Fluorescent

Ch E

Indoxyl Acetate

Non Fluorescent

O_2

VIII

Non Fluorescent

The acetate and butyrate esters of α- and β-naphthol
are hydrolyzed by cholinesterase to α-naphthol (λ_{ex} =
330 mμ; λ_{em} = 460-470 mμ) and β-naphthol (λ_{ex} = 320 mμ;
λ_{em} = 410 mμ). In general, the naphthol esters are not
as sensitive substrates as indoxyl acetate or resorufin
butyrate for the determination of cholinesterase, but
as little as 0.0005 units per ml could be detected with
an accuracy of about 3%.

Guilbault, Sadar, Glazer and Skou[14] prepared several
esters as substrates for cholinesterase: the acetate,
propionate and butyrate esters of N-methyl indoxyl,
umbelliferone and 4-methyl umbelliferone. Comparison
of these substrates with other fluorogenic esters:
indoxyl acetate, indoxyl butyrate, resorufin acetate,
β-carbonaphthoxy choline and β-naphthyl acetate, indicated
that N-methyl indoxyl acetate and butyrate were the best
substrates for true and pseudo cholinesterase, respectively. Analysis of as little as 5 x 10^{-5} units of
cholinesterase per ml can be performed
by a direct initial reaction rate method in 2-3 minutes
with an accuracy of about 1.5%.

Data on the comparison of various substrates for horse
serum cholinesterase is given in Figure 3 and Table 1.
All three N-methyl indoxyl esters are very stable in
solution (Figure 3), have a very low rate of spontaneous hydrolysis and a high rate of enzymic hydrolysis
(Table 1). All have low K_m values, and the N-methyl
indoxyl formed is not easily air oxidized to indigo
derivatives. The N-methyl derivatives also do not
appear to be as light sensitive as the corresponding
nor compounds, and stock solutions of these substrates
have been used for weeks with good results. The N-
methyl indoxyl formed is, as expected, as fluorescent
as indoxyl, but with shifted excitation and emission
wavelengths.

TABLE 1

Comparison of Various Substrates for
Horse Serum Cholinesterase

Substrate[a]	Blank[b]	Rate[c]	Km^d
Indoxyl Acetate	0.54	9.76×10^{-4}	3.4×10^{-4}
Indoxyl Butyrate	0.09	1.46×10^{-3}	--------
N-Methyl Indoxyl Acetate	0.035	1.765×10^{-3}	2.5×10^{-4}
N-Methyl Indoxyl Propionate	0.017	1.538×10^{-3}	1.17×10^{-4}
N-Methyl Indoxyl Butyrate	0.00	1.44×10^{-3}	1.5×10^{-4}
Resorufin Acetate	0.30	2.0×10^{-6}	8.0×10^{-5}
β-Carbonaphthoxy Choline	0.08	8.5×10^{-5}	--------
β-Naphthyl Acetate	0.01	1.4×10^{-4}	1.8×10^{-4}
Umbelliferyl Acetate	4.2	1.25×10^{-5}	--------
4-Methyl Umbelliferyl Acetate	4.3	1.25×10^{-5}	--------
4-Methyl Umbelliferyl Butyrate	6.8	2.7×10^{-5}	--------

[a]Optimum substrate concentration and pH used.

[b]Rate of spontaneous hydrolysis expressed as Δ fluorescence units per min.

[c]Rate in moles of substrate hydrolyzed per mg. of horse serum cholinesterase per minute. Blank rate subtracted. Rate calculated by dividing the observed rate in Δ fluorescence per minute by the total fluorescence per Molarity of each substrate.

[d]Michaelis constant for substrate - horse serum cholinesterase, obtained at optimum assay conditions described in reference (14).

Although the rate of cholinesterase hydrolysis of the indoxyl acetate is slightly higher than that of N-methyl indoxyl butyrate, the lower rate of non-enzymic hydrolysis of the latter permits the determination of lower concentrations of horse serum cholinesterase.

Results obtained on the hydrolysis of N-methyl

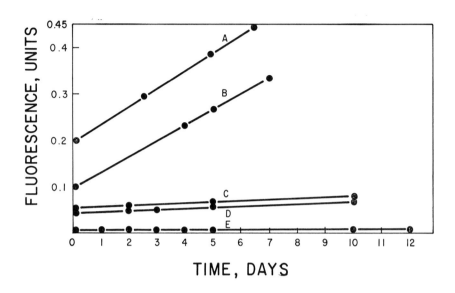

FIG. 3

Stability of various substrates in solution.
Stock solutions of all substrates were prepared in methyl cellosolve and their fluorescence measured each day after dilution
with buffer to optimum substrate and pH conditions (ref. 14).

    A - 4-Methyl umbelliferone butyrate
    B - Indoxyl butyrate
    C - N-methyl indoxyl butyrate
    D - N-methyl indoxyl propionate
    E - N-methyl indoxyl acetate

indoxyl acetate, propionate and butyrate by various
enzymes are given in Table 2.

TABLE 2

Effect of Various Enzymes on the Hydrolysis of
N-Methyl Indoxyl Esters

Enzyme[a]	Rate of Hydrolysis of Esters, $\Delta F/min$ [b]		
	Acetate	Propionate	Butyrate
Cholinesterase, Horse Serum	17.65	15.38	14.40
Cholinesterase, Bovine Erythro-cytes	2.9	1.03	0
Acid Phosphatase	0	1.1	1.1
Alkaline Phospha-tase	0	0.6	0.6
Lipase, Porcine Pancreas	2.1	4.4	3.8
Cellulase	1.2	0.7	0
β-Chymotrypsin	0	0	0
β-Glucosidase	0.28	0.12	0

[a]Concentration of all enzymes 0.03 mg./ml.

[b]Δ Fluorescence Intensity per minute.

Only horse serum, bovine erythrocyte cholinesterase
and lipase hydrolyze N-methyl indoxyl acetate at an
appreciable rate. Acetyl cholinesterase (bovine erythrocyte) does not catalyze the hydrolysis of the
butyrate ester; therefore, an assay of butyryl cholinesterase (horse serum) can be made in the presence of
acetyl cholinesterase using this ester with no interference from the latter. Lipase is the only serious
interference. Hydrolysis observed with phosphatase,
cellulase and glucosidase probably reflect the presence of other common esterases in these enzymes.

e. Radiometric Methods. Sensitive methods for assay of cholinesterase have been developed, based on radiometric methods. Procedures for acetylcholinesterase have been proposed by Winteringham and Disney[15], Reed, Goto and Wang[16] and Potter.[17] Acetyl-1-C^{14} choline is used as substrate and the concentration of liberated acetic acid-1-C^{14} is determined by counting. Unhydrolyzed substrate is removed by ion exchange and the radioactive acetic acid is counted. Assays can be performed in 30 minutes with accuracies of 3% or better.

3. Cellulase

Cellulase catalyzes the conversion of insoluble cellulose into soluble carbohydrates.

Cellulose (insoluble) ─────> soluble carbohydrate

Classical methods for cellulase have been based on one of three general characteristics of the enzyme: (a) the polymerized substrate (cellulose) has a high viscosity, which is reduced upon enzymic action[18]; (b) cellulose, being water insoluble, forms a colloidal suspension which scatters light. Cellulase degrades this substrate to form water soluble substances as glucose; such changes can be followed by a nephelometric technique[19] or (c) reducing sugars, such as glucose, are liberated on hydrolysis which can be detected by standard chemical or enzymic methods.[20] The first two methods are complex and imprecise (deviation of 10%). The third is more precise (5% deviation) but requires a long reaction time and is sometimes complicated by the supramolecular arrangement of the cellulose molecules within the substrate.

Guilbault and Heyn[21] tested several fluorogenic substrates for cellulase, namely fluorescein dibutyrate, α- and β-naphthyl acetate, indoxyl acetate and resorufin butyrate (Table 3). Of all the esters tried, only

TABLE 3

Hydrolysis of Various Fluorogenic Substrates by
Cellulase

Cellulase concentration = 0.030 units/ml.;
pH and substrate concentrations optimum

Substrate	Type Enzyme	ΔF/min.[a]	Blank[b]
α-Naphthyl acetate	Worthington CSE-1	0.2	0
β-Naphthyl acetate	Worthington CSE-1	0.3	0
Fluorescein dibutyrate	Worthington CSE-1	0.0	0.001
Indoxyl acetate	Worthington CES-1	2.5	0.33
Resorufin acetate	Worthington CSE-1	36.5	0.1
	Rohm and Haas-36	2.9	0.1
	Wallerstein	2.7	0.1

[a]Rate of change in fluorescence intensity with time.

[b]Rate of formation of fluorescent product with no enzyme added.

indoxyl acetate and resorufin acetate were hydrolyzed
at an appreciable rate. The latter is cleaved by
cellulase to give the highly fluorescent resorufin(V)
(λ_{ex} = 540 mµ; λ_{em} = 580 mµ). Using this substrate
from 0.00010 to 0.060 units per ml of cellulase can be
determined with a precision of about 1.5% in 1-2

$$\text{Resorufin Acetate} \xrightarrow{\text{Cellulase}} \text{Resorufin}$$
$$\text{(Non-Fluorescent)} \qquad\qquad \text{(Fluorescent)}$$

minutes (Table 4). This proposed method represents
a considerable improvement over any other method,
not only in the reduction of analysis time, but in
precision and sensitivity.

TABLE 4
Results of Determination of Cellulase
Resorufin acetate=$3 \times 10^{-5}\underline{M}$, $0.01\underline{M}$ tris buffer, pH 7.0

Cellulase, units/ml.			Rel. Error, %	
Present[a]	Found[b]	Found[c]	Fl.Method	Color.Method
0.000200	0.000203	0.000190	+ 1.5	- 5.0
0.00100	0.00102	0.00106	+ 2.0	+ 6.0
0.00300	0.00295	0.00290	- 1.7	- 3.3
0.0120	0.0122	0.0125	+ 1.6	+ 4.0
0.0300	0.0300	0.0285	0.0	- 5.0
0.0600	0.0590	0.0628	- 1.7	+ 4.7
Av. Rel. Error			± 1.5	± 4.7

[a]Amount of enzyme added.

[b]Amount found by fluorescence (fl) method. Each result represents an average of three or more determinations with a relative standard deviation of 1.5%.

[c]Amount found by colorimetric assay of liberated glucose using glucose oxidase, peroxidase and o-dianisidine.

Resorufin acetate, although an ideal substrate for cellulase, does have an appreciable rate of spontaneous hydrolysis, particularly at a pH $\rangle$ 7.5-8 and at high ionic strength. Working at a pH of 7 with an ionic strength (μ) of $0.01\underline{M}$, a very low spontaneous rate of hydrolysis was observed. The substrate stock solution, $10^{-2}\underline{M}$ in methyl cellosolve, is stable for several months in a refrigerator with very little fluorescence observed.

Evidence for cellulase, and not esterase, activity was (a) that the enzymic activity measured with resorufin acetate was inhibited by known, specific inhibitors of cellulase and (b) the activity measured fluorometrically parallels the activity found by another standard procedure (Table 4).

4. Chrymotrypsin

Chrymotrypsin is one of the major proteolytic enzymes. Although, it acts on a wide variety of peptide and ester linkages it preferentially hydrolyzes bonds involving L-tyrosine or L-phenylalanine. The D-isomers are not attacked. Desnuelle has published a lengthy review on chymotrypsin.[22]

Several colorimetric methods have been proposed for the assay of chymotrypsin. An acylating agent, N-trans cinnamoylamidazole, was used by Schonbaum, Zerner and Bender[23] for the spectrophotometric titration of chymotrypsin. The reagent reacts with the active site of the enzyme and gives an absolute measure of its concentration. Erlanger and Edel[24] used 2-nitro-4-carboxyphenyl-N-N-diphenyl carbamate for the direct determination of chymotrypsin activity. The compound reacts with the enzyme to produce diphenyl carbamyl-chymotrypsin plus 3-nitro-4-hydroxybenzoic acid which is yellow.

Several chromogenic substrates have been prepared for chymotrypsin: N-carbobenzoxy L-tyrosine p-nitro-phenyl ester[25], N-benzoyl-L-tyrosine-p-nitroanilide,[26] acetyl-L-tyrosine ethyl ester (ATEE)[27], and benzoyl-L-tyrosine ethyl ester (BTEE).[28] ATEE has been widely used for the analysis of trypsin as well as chymotrypsin. For assay of chymotrypsin, BTEE is the most useful substrate, since it is completely resistant to hydrolysis by trypsin. In this procedure the rate of hydrolysis of BTEE is determined from the change in absorbance at 256 mμ. One unit is equivalent to one μmole of substrate hydrolyzed per minute at pH 7.8 and 25°C.

Fluorometric methods for the assay of chymotrypsin have been proposed by Bielski and Freed[29] and Guilbault and Kramer.[30] Bielski and Freed used N-acetyl-L-tryptophan ethyl ester and N-acetyl-L-tyrosine ethyl ester as substrates for chymotrypsin. The rate of production of the fluorescence of tryptophan or tyrosine indicated the amount of enzyme present.

Guilbault and Kramer[30] proposed the use of fluorescein dibutyrate as a substrate for chymotrypsin. The method was based on the hydrolysis of this non-fluorescent ester by chymotrypsin. The rate of change in the fluorescence of the solution due to production of fluorescein, $\Delta F/min$, is measured and correlated with enzyme activity. From 0.167 to 1.30 mg per ml of α-,

$$\text{Fluorescein dibutyrate} \xrightarrow[\text{pH 7.4}]{\text{chymotrypsin}} \text{Fluorescein}$$
$$\text{(non-fluorescent)} \qquad (\lambda_{ex} = 490 \text{ m}\mu;$$
$$\lambda_{em} = 520 \text{ m}\mu)$$

β- or γ-chymotrypsin could be determined in 1-2 minutes with a relative standard deviation of about 2%. Lipase is an interference.

The fluorescein dibutyrate is a fairly stable substrate. A $10^{-3}\underline{M}$ stock solution of this substrate is generally prepared in methyl cellosolve and stored under refrigeration. Dilutions with buffer are made before each determination.

5. Beta-Glucosidase and Galactosidase

β-Glucosidase catalyzes the hydrolysis of alkyl and aryl β-D-glucoside:

$$\beta\text{-D-glucoside} + H_2O \longrightarrow \text{D-Glucose} + \text{alcohol}$$

Sweet almond is a rich source of this enzyme, which is found widely distributed among plants. Substrates include salicin, amygdalin, cellobiose and gentiobiose. Reviews on β-glucosidases have been compiled by Larner[31] and Veibel.[32]

In the spectrophotometric method of Nelson[33] salicin is cleaved to give saligenin + β-D-glucose. The glucose produced is then assayed colorimetrically. A

ENZYMATIC METHODS OF ANALYSIS

unit of activity is that liberating one micromole of glucose per minute under specified conditions. Alternatively, arbutin (β-D-glucosido-hydroquinone) can be used as substrate; the glucose produced can be assayed iodometrically.[34] Of the two substrates salicin is undoubtedly the choice. With this substrate analysis can be performed in about 15 minutes, compared to 97 hours with arbutin.

Guilbault and Kramer[35] have devised a simple electrochemical reaction rate method for the analysis of glucosidase. The method is based on the liberation of cyanide from the substrate amygdalin(IX); the cyanide produced is measured using a silver-platinum electrode combination in a spontaneous (internal) electrolysis cell.

$$2 CN^- + Ag^0 \longrightarrow Ag(CN)_2^- + e^-$$

Before addition of glucosidase, the potential of the system is approximately zero. Upon enzymolysis a change in potential results due to the reaction of the cyanide produced with the silver electrode (Fig. 4). The rate of change in the potential of the system with time is proportional to the glucosidase concentration over the range 0.00156 to 0.078 unit/ml of total solution with a relative standard deviation of about 2%.

Umbelliferone (7-hydroxycoumarin) and 4-methylumbelliferone are highly fluorescent compounds which have been modified to form non-fluorescent substrates for the enzymes glucuronidase, glucosidase and galactosidase. Robinson[36] used 4-methylumbelliferone β-D-glucoside

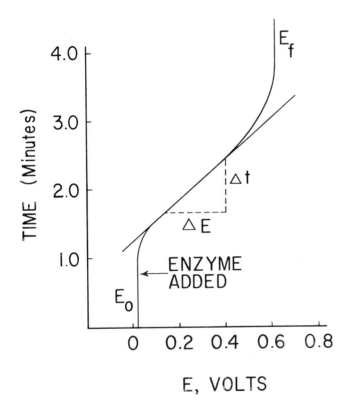

FIG. 4

Change in potential of a 0.005 M
amygdalin solution upon addition
of 0.02 mg. of glucosidase (ref. 35).

as a substrate for β-glucosidase. The substrate is split
specifically by this enzyme to the highly fluorescent
4-methylumbelliferone.

Woolen and Walker[37] proposed 4-methylumbelliferyl
β-D-galactoside as a substrate for β-galactosidase, and
Rotman, Zderic and Edelstein[38] determined this enzyme
fluorometrically in a similar manner.

6. Beta-Glucuronidase

β-Glucuronidase catalyzes the hydrolysis of conjugated
β-glucuronides to D-glucuronic acid and the corresponding
alcohol. Typical substrates for the enzyme include the
β-glucuronides of phenolphthalein, 8-hydroxy quinoline,
chlorophenol, naphthol, etc. In the colorimetric assay
of the activity of β-glucuronidase, phenolphthalein
glucuronide is used as substrate. The phenolphthalein
liberated is measured by its red color at alkaline pH.
The initial rate of change in absorbance with time, $\Delta A/$
min., is proportional to the substrate concentration as
well as to the concentration of β-glucuronidase. A
unit of β-glucuronidase is that cleaving one micromole
of phenolphthalein glucuronide per minute at $37°C.$[39]

Woolen and Turner[40,41] and Mead et al[42] have
proposed the use of the β-glucuronides of umbelliferone
and 4-methylumbelliferone as substrates for β-glucuronidase. The rate of production of the fluorescent
umbelliferone alcohols is a measure of the activity of
the enzyme. About 2 orders of magnitude increase in
sensitivity over the colorimetric procedure is obtained.

Veritz, Caper and Brown[43] and Greenberg[44] assayed
β-glucuronidase using 2-naphthyl-β-D-glucuronide as
substrate. The reaction produces the fluorescent
β-naphthol and can be followed kinetically at pH of 5.3,
or more sensitively by making the solution alkaline
(pH 13) prior to measurement of the fluorescence.

7. Hyaluronidase

Testicular hyaluronidase hydrolyzes the endo-N-acetyl-hexosaminic bonds of hyaluronic acid to tetrasaccharide residues. It is a glycosidase with hydrolytic and transglycosidase activity.[45,46]

Classical methods for hyaluronidase are based upon one of three general characteristics: (1) the poly-merized substrate (hyaluronic acid) forms salt linkages with proteins to give acid-insoluble complexes except when it is enzymatically depolymerized[47,48]; (2) the substrate has a high viscosity when dissolved in solu-tions of low ionic strength, the viscosity is reduced upon enzymic action[49,50] or (3) upon hydrolysis re-ducing sugars are liberated that can be detected by standard methods.[51] All of these methods leave much to be desired from an analytical point of view. The turbidimetric method is complex and inaccurate (10%); the viscosity reduction technique is fairly accurate, but is very tedious and time consuming.

Guilbault and Kramer[52] have described a simple, rapid fluorometric assay of the enzyme hyaluronidase based upon the hydrolysis of nonfluorescent indoxyl acetate by the enzyme to give the highly fluorescent indigo white (VIII). By this procedure 0.0010 to 0.033 µg/ml of hyaluronidase can be assayed in 1-2 minutes with precision of about 1.8%. In order to ascertain whether true hyaluronidase or simple esterase activity was being measured, the enzyme was tested for inhibition by various inorganic and organic compounds. Hyaluronidase is known to be inhibited by Fe^{2+} and Cu^{2+} and is relatively insensitive to organophosphorus compounds, whereas esterases are strongly inhibited by organophosphorus compounds. Heavy metal ions such as Pb^{2+}, Ag^+ and Hg^{2+} have been shown to strongly inhibit esterases, but not hyaluronidase.

Experimentally it was found that Pb^{2+}, Ag^+ and Hg^{2+} had no significant effect on the rate of hydrolysis, but Fe^{2+} and Cu^{2+} strongly inhibited the enzymic activity. Little effect, likewise, was noted from organophosphorus compounds, thus proving that indoxyl acetate is being cleaved by hyaluronidase and not esterase, and indoxyl acetate is thus an excellent substrate for the assay of hyaluronidase activity. Esterase activity can be masked by addition of an organophosphorus compound, like systox, allowing the assay of hyaluronidase.

8. Lipase

Lipase is an esterase that cleaves large molecular substrates, i.e. triglycerides. Originally, it was believed that activation by bile salts was necessary for lipolytic activity, but it has been recently shown that taurocholate is not a reliable activator.[53,54] Guilbault and Kramer[53] have shown that taurocholate is an activator only when an insoluble interface exists between the substrate and enzyme, and taurocholate is not necessary when soluble substrates are used.

Several colorimetric substrates have been proposed for assay of lipolytic activity. Of the phenol and naphthol esters proposed as substrates, 2-naphthyl nonanoate is the most sensitive. Unfortunately, this substrate is also cleaved by other esterases, so is not specific. Ravin and Seligman[55] proposed the use of the esterase resistant 2-naphthyl myristate as substrate for assay of lipase in the diagnosis of human pancreatitis.[54]

In this method the myristate ester is hydrolyzed to 2-naphthol. Two molecules of 2-naphthol are coupled with tetrazotized o-dianisidine to give a purple azo dye (λ_{max} = 540 mμ) which is determined colorimetrically. A five hour incubation time was found necessary for reliable diagnosis of pancreatitis. Cholate was added

to suppress esterase, and fully activate lipase activity.

In recent publications, Kramer and Guilbault described a simple, rapid procedure for the assay of
lipase activity in the presence of other esterases,
based on the hydrolysis of fluorescein esters catalyzed
by lipase.[56,57] By following the rate of production
of the fluorescence of fluorescein with time, a series
of curves are obtained, the slopes of which, $\Delta F/\Delta t$,
are proportional to the lipase concentration over the
range 10^{-3} to 10^{-1} units per ml. The method is rapid
(1-2 minutes for analysis) and accurate(1%).

$$\text{Dibutyrylfluorescein} \xrightarrow{\text{lipase}} \text{fluorescein}$$
$$\text{(non-fluorescent)} \qquad (\lambda_{ex} = 490 \text{ m}\mu;$$
$$\lambda_{em} = 520 \text{ m}\mu)$$

Esterases were found to have little effect on this
substrate, thus providing the specificity desired in
a lipase analysis. Bile salts were found to be unneces-
sary since the substrate and enzyme are both soluble
and form a homogenous solution upon mixing.

Sapira and Shapiro[58] used fluorescein dibutyrate
for the assay of hormone insensitive lipase of rat
adipose tissue, and found no interference from common
esterases. Parkin[59] has suggested the use of esters
of fluorescein for the development of methods for the
assay of human sera lipase.

Jacks and Kircher[60] synthesized and tested the
butyryl, hexanoyl, heptanoyl, nonanoyl, palmitoyl and
oleoyl esters of 4-methylumbelliferone as substrates
for 5 preparations of lipase. Enzymic hydrolysis re-
leased the intensely fluorescent 4-methylumbelliferone.
The highest rate of hydrolysis was obtained with the
hexanoyl ester for steapsin, the heptanoyl ester for
wheat germ and peanut lipases, the octanoyl ester for

castor bean lipase and the nonanoyl ester for porcine
pancreas lipase. It was claimed that the rate of
hydrolysis of these esters proceeded at a higher rate
than the dibutyryl esters proposed by Guilbault and
Kramer[56,57], but no data was given on the lowest
detectable concentration of enzymes.

In a complete study of fluorogenic substrates for
lipase, Guilbault and Sadar[61] evaluated 12 different
compounds: fluorescein dibutyrate from both Eastman
Organics and Nutritional Biochemical Co., umbelliferyl
acetate, 4-methylumbelliferyl-acetate, - butyrate,
-heptanoate, -octanoate, -nonanoate, and -caproate and
the acetate, proprionate and butyrate esters of N-methyl
indoxyl.[14]   All substrates were compared with respect to stability, spontaneous hydrolysis, enzymic
hydrolysis, Michaelis constant for the enzyme-substrate
complex and total fluorescence of final product. Optimum conditions of analysis were found for all substrates
and using these conditions, the lowest detectable enzyme
concentration was found for each substrate. The results
of this study are summarized in Table 6 and Fig. 5.

The blank rates were determined by recording the
change in fluorescence with time over a period of 5
minutes. Then 0.1 ml of a stock 1 mg/ml solution of
porcine pancreas was added, and the rate of enzymic
cleavage determined by dividing the observed rate,
ΔF/min., by the fluorescence coefficient (fluorescence
of the product formed, i.e. umbelliferone, divided by
the concentration of fluorescent product in $\underline{M}$). The
fluorescence coefficient can be determined by measuring
the fluorescence of a known concentration of the product. The lowest detectable concentration of lipase
reported is that concentration required to give an
enzymic rate twice that of the blank rate.

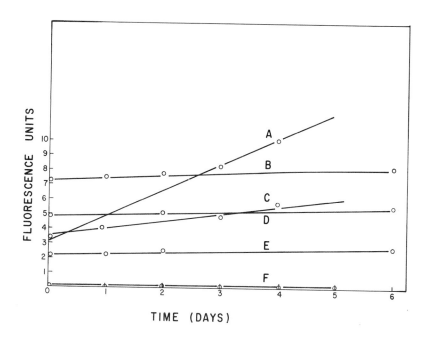

FIG. 5

Stability of Various Substrates in Solution

   Stock solutions of all substrates were
prepared in methyl cellosolve and their
fluorescence measured each day after
dilution with buffer to optimum substrate
and pH conditions.

   A - Fluorescein Dibutyrate (Eastman)
   B - 4-Methylumbelliferone Caproate
   C - Fluorescein Dibutyrate (NBC)
   D - 4-Methylumbelliferone Octanoate
   E - 4-Methylumbelliferone Butyrate
   F - N-Methyl Indoxyl Butyrate

TABLE 6

Comparison of Various Substrates for Lipase

Substrate	Fluorescence Wavelengths	Optimum Conditions[a]	Blank[b]	Rate[c]	Lowest Detectable Conc. (mg./mL.)	K_m
Fluorescein Dibutyrate (NBC)	$\lambda ex=490$ mμ $\lambda em=520$ mμ	S=1.56×10^{-4} B=0.1$\underline{M}$, pH7.5	0.080	1.7×10^{-6}	0.0040	7×10^{-6}
N-Methyl Indoxyl Acetate	$\lambda ex=430$ mμ $\lambda em=500$ mμ	S=4.66×10^{-4} B=0.1$\underline{M}$, pH7.5	0.037	1.0×10^{-3}	0.00030	-----
N-Methyl Indoxyl Propionate	Same	Same	0.027	1.17×10^{-3}	0.00025	-----
N-Methyl Indoxyl Butyrate	Same	Same	0.014	1.4×10^{-3}	0.00020	2.9×10^{-5}
Umbelliferone Acetate	$\lambda ex=330$ mμ $\lambda em=450$ mμ	S=3.1×10^{-5} B=0.1$\underline{M}$, pH7.0	0.04	3×10^{-7}	0.010	-----
4-Methylumbelliferone Acetate	Same	S=3.1×10^{-5} B=0.1$\underline{M}$, pH6.5	0.01	1.6×10^{-6}	0.00280	-----
4-Methylumbelliferone Butyrate	Same	S=3.1×10^{-5} B=0.1$\underline{M}$, pH7.0	0.02	1.1×10^{-5}	0.00040	-----
4-Methylumbelliferone Heptanoate	Same	S=3.1×10^{-5} B=0.1$\underline{M}$, pH6.5	0.01	9.6×10^{-5}	0.000035	7.3×10^{-6}

TABLE 6 (Continued)

4-Methylumbelliferone Octanoate	Same	S=3.1x10^{-5} B=0.1$\underline{M}$,pH6.5	0.01	8.3x10^{-5}	0.000050	8.0x10^{-6}
4-Methylumbelliferone Nonanoate	Same	S=3.1x10^{-5} B=0.1$\underline{M}$,pH6.5	0.01	7.6x10^{-5}	0.000080	-----
4-Methylumbelliferone Capreate	λ_{ex}=340 mμ λ_{em}=450 mμ	S=3.1x10^{-5} B=0.1$\underline{M}$,pH7.5	0.01	5.9x10^{-5}	0.00010	-----

[a]S=Substrate Concentration in $\underline{M}$, B=Phosphate Buffer

[b]Rate of Non-Enzymic Hydrolysis expressed in Δ Fluorescence Units per min.

[c]Rate of $\underline{M}$ of substrate hydrolyzed per mg. of porcine pancreas lipase per minute. Blank rate subtracted. Rate calculated by dividing the observed rate in Δ fluorescence per minute by the total fluorescence per $\underline{M}$ of each substrate.

From all aspects, 4-methylumbelliferone heptanoate
(X, $R=C_7H_{15}$) was found to be the best substrate for

R-C-O   X $\xrightarrow{\text{Lipase}}$ HO   XI

X   XI
Non-Fluorescent   Fluorescent
                           (λ_{ex} = 330 mμ;
                             λ_{em} = 450 mμ)

porcine pancreas lipase, and 4-methylumbelliferone
octanoate (X, $R=C_8H_{17}$) was best for fungal lipase. As
little as 2.0 x 10^{-5} unit per ml of solution could
be determined by a direct reaction rate method in 2-3
minutes with a precision of about 1.5%.

The 4-methylumbelliferone heptanoate and octanoate
esters were found to be stable for several weeks with
little increase in fluorescence. This compares favor-
ably with other substrates.

The effect of various enzymes on the hydrolysis of
4-methylumbelliferyl heptanoate under the optimum conditions described in Table 6 is given in Table 7. Por-

TABLE 7

Effect of Various Enzymes on the Hydrolysis
of 4-Methylumbelliferyl Heptanoate

Enzyme (0.031 mg/ml)	$\Delta F/min$[a]
Lipase, Porcine Pancreas	32.0
Cholinesterase, Horse Serum	4.8
Cholinesterase, Bovine Erythrocyte	0
Cellulase	0.49
β-Chymotrypsin	2.4
β-Glucosidase	0

[a]Rate of change in fluorescence intensity with time.

cine pancreas lipase, horse serum cholinesterase, β-chymotrypsin and acid phosphatase all are capable of effecting the hydrolysis of this ester.

9. Phosphatase

Phosphatases are enzymes that catalyze the hydrolysis of phohsphate esters. They are classified as "acid" or "alkaline" depending on their pH optimum (5-6 for acid phosphatase, 8-10 for alkaline phosphatase).

Several substrates have been proposed for the colorimetric assay of acid and alkaline phosphatase. Phenolphthalein diphosphate, proposed by Huggins and Talalay[62] and Linhard and Walter[63] is hydrolyzed by phosphatase to phenolphthalein which is determined colorimetrically at 530 mμ. Both phosphate groups must be split off

Phenolphthalein Diphosphate ————>

$\qquad$ Phenolphthalein monophosphate

$\qquad\qquad$ ↓

$\qquad\qquad$ phenolphthalein

$\qquad\qquad$ (λ_{max} = 530 mμ)

before the phenolphthalein color can be formed in alkali; hence phenolphthalein monophosphate has been described as a better substrate for phosphatase.[64] Similarly Coleman[65] used thymolphthalein monophosphate as a substrate.

Several workers[66-69] have proposed the use of p-nitrophenyl phosphate as substrate. The rate of p-nitrophenol formation at 405 mμ is measured and is proportional to the phosphatase activity. Phenyl phosphate has been proposed as a substrate for the determination of phosphatase in milk.[70,71] After a two hour incubation the phenol liberated is determined colorimetrically with 2,6-dibromoquinone-4-chlorimide at 600 mμ.

Because of the greater sensitivity of fluorometric methods, several procedures have been described using substrates that are cleaved to yield fluorescence

products. Moss[72] has prepared naphthyl phosphates
as substrates for both acid and alkaline phosphatase.
The rate of formation of the fluorescent α-naphthol
(liberated when phosphatase is added to α-naphthyl
phosphate) is recorded, and the amount of the enzyme
calculated from calibration plots of the initial rate
vs. the concentration of phosphatase. Greenberg[73]
used sodium β-naphthyl phosphate as a substrate for
alkaline phosphatase, in a procedure similar to that
of Moss. Takeuchi and Nogami[74] used riboflavin-5'
phosphate as substrate in the histochemical measurement of tissue phosphatase. Elevitch et al.[75] have
described a film badge technique for phosphatase isoenzymes using β-naphthyl phosphate or 3-0-methyl
fluorescein phosphate as substrate.

Land and Jackim[76] have found flavone-3-diphosphate
to be a stable, versatile and sensitive substrate for
assaying both acid and alkaline phosphatase. The fluorescence of 3-hydroxyflavone is measured at 510 mμ.
The authors found this substrate to be more sensitive
than β-naphthyl phosphate and more stable than 3-0-
methylfluorescein phosphate. Better sensitivity was
claimed by forming the fluorescent metal chelate of
3-hydroxyflavone with aluminum ions.

Fernley and Walker[77] investigated 4-methyl-
umbelliferyl phosphate as a substrate for calf
intestinal alkaline phosphatase. Kinetic and thermodynamic constants were determined.

Guilbault, Sadar, Glazer and Haynes[78] prepared
umbelliferone phosphate(XI) as a substrate for acid
and alkaline phosphatase, and compared this substrate
with other substrates described in the literature.
These results are summarized in Tables 8 and 9. The
two colorimetric substrates are relatively insensitive,
having low hydrolysis rates, and require large samples.
Some improvement is observed with the naphthyl esters,

TABLE 8

Comparison of Various Substrates for Acid Phosphatase

Substrate	Blank[a]	Rate[b]	Sample Size ml. of Serum	Method of Measurement	Conc. of PO_4^{3-} for 50% Inhibition, M
p-Nitrophenyl Phosphate[c]	0	0.09	0.1	Colorimetrically	1.2×10^{-4}
Phenolphthalein Phosphate	0	0.10	0.1	"	------
α-Naphthyl Phosphate	0	0.8	0.05	Fluorometrically	------
β-Naphthyl Acetate	0.017	0.76	0.05	"	------
β-Naphthyl Phosphate	0	0.88	0.05	"	------
Indoxyl Acetate	0.20	1.16	0.05	"	8.6×10^{-3}
3-Hydroxyflavone Phosphate	0	2.0	0.005	"	------
Umbelliferone Phosphate	0	7.74	0.001	"	2.7×10^{-4}

[a] Rate of spontaneous hydrolysis expressed in Δ Absorbance or Δ Fluorescence Units per min.

[b] Rate with 0.03 mg./ml. of Type II Acid Phosphatase, in Δ A or Δ F/minute.

[c] di-Tris Salt.

TABLE 9

Fluorescence Coefficients and Michaelis Constants
For Various Substrates

Substrate	Km, Phosphatase Acid	Km, Phosphatase Alkaline	Fluorescence Coefficient[a]
Umbelliferone Phosphate	5×10^{-5}	1.9×10^{-5}	2.7×10^{6}
4-Methylumbelli- ferone Phosphate	-------	4.4×10^{-5}[b]	2.2×10^{6}
3-Hydroxy Flavone Phosphate	-------	8.8×10^{-6}[c]	1.8×10^{5}
β-Naphthyl Phosphate	-------	1.7×10^{-4}[d]	1.5×10^{5}[d]
α-Naphthyl Phosphate	-------	----------	5×10^{4}[d]
Indoxyl Acetate	-------	----------	1×10^{5}

[a]Fluorescence of Hydrolyzed Substrate divided by the
concentration of original in M.   Value for Quinine
Sulfate in 0.1 N sulfuric acid at a λ_{ex} = 350 mμ and
a λ_{em} = 450 mμ was 1.75×10^{6} mole^{-1}.

[b]Reference 77, Ionic Strength 0.10 pH 9.24.

[c]Reference 76.

[d]Reference 72.

phosphate and acetate, but the hydrolysis products from
these substrates, α- and β-naphthol, are not extremely
fluorescent (Table 9).   Thus an increase in sensitivity
over colorimetric substrates is achieved, but this
increase is not large.   Indoxyl acetate, reported to be
a good substrate for cholinesterase[12], has an appreciable rate of enzymic hydrolysis and yields a highly
fluorescent product.   A high rate of spontaneous hydrolysis at the pH values required for the phosphatase
assay makes this a poor substrate.

From aspects of stability, rate of enzymic hydrolysis
and fluorescence of product formed, umbelliferone phos-
phate(XI) appears to be an ideal substrate. The umbelli-
ferone(X) formed has a higher fluorescence than the
naphthols, indoxyl, 3-hydroxyflavone and 4-methylumbelli-
ferone. The K_m value is better than all substrates except 3-hydroxyflavone phosphate. Using umbelliferone
phosphate as little as 10^{-6} unit of alkaline phosphatase
and 10^{-5} unit of acid phosphatase may be detected and
measured by a direct initial rate method in 2-3 minutes.
Concentrations ranging from 10^{-6} to 2×10^{-2} units per
ml of alkaline phosphatase and 10^{-5} to 0.06 units per ml
of acid phosphatase were assayed with an error and preci-

(Non-Fluorescent)   (Fluorescent)

XI   X

sion of about 1.5%.

The solution of umbelliferone phosphate was found to
be stable for at least a week with little increase in
fluorescence produced, and with proper case, solutions
should be stable for months.

The effect of various enzymes on the hydrolysis of
umbelliferone phosphate at pH values of 5 and 8 is given
in Table 10. Only phosphatase and β-glucosidase have
any appreciable effect on the substrate; the rate of
hydrolysis with cellulase, lipase, horse serum and
bovine erythrocyte cholinesterases and chymotrypsin is
zero. Thus this substrate appears to be an excellent
one for the determination of acid phosphatase at pH 5
and alkaline phosphatase at pH 8.

TABLE 10

Effect of Various Enzymes on the Hydrolysis
of Umbelliferone Phosphate

Enzyme[a]	pH	ΔF/min.	pH	ΔF/min.
Acid Phosphatase	5	22.2	8	0.48
Alkaline Phosphatase	5	2.2	8	65.3
Cellulase	5	0	8	0
Cholinesterase, Horse Serum	5	0	8	0
Acetylcholinesterase, Bovine	5	0	8	0
Lipase, Porcine Pancreas	5	0	8	0
Chymotrypsin	5	0	8	0
β-Glucosidase	5	1.63	8	0.21

[a]Concentration of all enzymes 0.03 mg./ml.

10. Urease

Urease catalyzes the hydrolysis of urea:

$$NH_2 - \underset{O}{\overset{\|}{C}} - NH_2 + H_2O \rightleftharpoons CO_2 + 2\ NH_3$$

This was the first enzyme to be isolated in crystal-
line form[79] and is one of the most selective ones.
Urease is most commonly determined by an acid-base
titration or by a Nessler determination of the NH_3
produced.   One unit of activity represents one
micromole of ammonia liberated per minute at $25^\circ C$ under
assay conditions.[80]

A radiometric procedure for urease was described by
Schatalova and Meerov.[81]   Carbon-14 labelled urea was
used, and the radioactive CO_2 gas was measured by standard counting procedures.

Katz[82] and Katz and Cowans[83] have described a
direct potentiometric method for urease by continuous
recording of the ammonium ion produced from urea using a
Beckman cation sensitive glass electrode.   Purdy,

Christian and Knoblock[84] described a method for
the analysis of urease based on the hydrolysis of urea
to form ammonia. The resulting ammonia is then titrated with coulometrically generated hypobromite using
an amperometric end point. In urine the ammonia
could be titrated directly; but in blood samples,
the ammonia is first separated by a microdiffusion
technique.

Urease activity can be determined colorimetrically[85] by reaction of the ammonium ion formed with
sodium hypochlorite and sodium phenolate solutions.
The blue indophenol formed is measured at 630 or 580
mμ after maximum color develops (about 20 minutes).
Linearity between absorbance and concentration is
obtained up to 100 μg of ammonia.

C.   OXIDATIVE ENZYMES

1.   Amino Acid Oxidases

D- and L-amino acid oxidases catalyze the oxidative deamination of D- and L-amino acids according
to the overall reaction:

$$R\text{-}CH\text{-}COOH + O_2 + H_2O \xrightarrow{\text{Enzyme}} R\text{-}C\text{-}COOH + NH_3 + H_2O_2$$
$$\underset{NH_2}{|} \qquad\qquad\qquad\qquad \overset{||}{O}$$

The D-amino acid oxidase reacts specifically with D-
amino acids; the L-oxidase only with L-amino acids.

Methods for the analysis of these oxidases fall
into one of three classes: (a) manometric methods,
in which the uptake of oxygen is measured; (b) electrochemical procedures in which the ammonia is
assayed and (c) colorimetric and fluorometric procedures involving measurement of the peroxide liberated.

In the manometric procedure[86] the reactants are
placed in a standard single-arm Warburg flask and the
oxygen uptake is read at 5 minute intervals for 30
minutes. A unit of activity is that which causes the

deamination of 1 micromole of L-leucine per minute
under specified conditions at $37^\circ C$.

Guilbault[87] has described an electrochemical method
for the assay of amino acid oxidase. An ammonium selective electrode is used to monitor the ammonium ion
liberated in the oxidation of amino acids; plots of
$\Delta E/\Delta t$ are proportional to the concentration of enzyme
over the concentration range of 10^{-3} to 10^{-1} unit.

Hydrogen peroxide is one of the reaction products when
an amino acid is deaminated by an amino acid oxidase.
By coupling a peroxidase-acceptor indicator reaction
with the amino acid oxidase reaction a highly specific
microdetermination of certain amino acids is possible.
Malmstadt and Hadjiioannou[88] proposed the use of
o-dianisidine as the indicator dye and developed a
sensitive method for the colorimetric determination of
D- and L-amino acid oxidases.

Guilbault and Hieserman[89] described a fluorometric
assay procedure for D- and L-amino acid oxidase based
on the conversion of the non-fluorescent homovanillic
acid(XII) to the highly fluorescent 2,2-dihydroxy-3,3-
dimethoxy-diphenyl-5,5' diacetic acid(XIII). The
initial rate of formation of this fluorescent compound
is measured and related to the activity of the enzymes
in the concentration range 0.00009 to 0.025 unit per
ml. Complete analysis can be performed in 2 minutes

$$\text{Amino Acid} \xrightarrow{\text{Oxidase}} H_2O_2$$

(Non-Fluorescent)   (Fluorescent)

XII   XIII

with a precision of 1.5% and an increase in sensitivity
of two orders of magnitude over other available procedures.

2. Catalase

Catalase catalyzes the following reaction:

$$2\ H_2O_2 \longrightarrow 2\ H_2O + O_2$$

There are many assays available for catalase and Maehly
and Chance have published a review of available
methods.[90,91] Some common methods for the assay of
catalase activity are:

a) Measurements of the oxygen liberated on decompo-
     sition of H_2O_2;
b) Determination of hydrogen peroxide by its
     absorbance of 240 mμ.;
c) Electrochemical measurement of hydrogen peroxide;
d) Titration of residual hydrogen peroxide volu-
     metrically.

Both manometric[92,93] and "Katalaser" volumetric[94]
procedures have been described for measurement of oxygen
liberation. The "Katalaser" is a graduated glass vessel
for the volumetric measurement of oxygen. It is not as
accurate as the Warburg manometer, but has the advantage
of simplicity.

The spectrophotometric procedure was first described
by Beers and Sizer[95] and is based on a measurement of
the absorbance of a hydrogen peroxide solution between
230-250 mμ. Catalase decomposes hydrogen peroxide and
the absorbance decreases with time. From measurements
of the change in absorbance with time, the activity of
catalase can be determined. Any substances that absorb strongly at these wavelengths will interfere.

Residual hydrogen peroxide can also be determined
by oxidation with ceric sulfate, potassium perman-
ganate, or iodine. Samples are removed from the

reaction flask at timed intervals, the reaction is
stopped with acid, and the peroxide determined by one
of the three methods. In the most commonly used iodi-
metric procedure the hydrogen peroxide not decomposed
by catalase oxidizes iodide to iodine. The liberated
iodine is titrated with thiosulfate using starch as
indicator.[96]

Guilbault[97] described an electrochemical method
for the analysis of the enzymes peroxidase and catalase.
The method is an adaptation of that described above for
the analysis of cholinesterase, in which a small constant current is applied across two platinum electrodes.
The enzyme catalase catalyzes the oxidation of hydrogen
peroxide to oxygen and water. Hydrogen peroxide is
electroactive, but the products of its oxidation are
not. In the electrochemical method proposed for cholinesterase a more easily oxidizable material was produced
upon enzymic hydrolysis, and hence the platinum electrode
quickly assumed the new potential, yielding a large
change of potential with time. However, in the case of
a change from a more easily oxidizable to a more difficult oxidizable species, as in the case of the conversion of hydrogen peroxide to oxygen and water, all of
the more electroactive species must be removed before
any change in potential will be observed. This can be
done experimentally by properly controlling the current
to a valve near the i_{max}, so that a small change in the
peroxide concentration will allow the voltage to increase until the potential of the solvent is reached
(Fig. 6). Using a current of $2\mu a$ applied across two
thimble electrodes immersed in a solution of peroxide,
a smooth initial potential of about 0.4v is obtained
(Fig. 7). Upon addition of catalase, enough of the
peroxide is removed at $2\mu a$ to allow the potential to
increase, until finally a potential of 1.0 volts is
reached (Figures 6 and 7). The polarographic curve

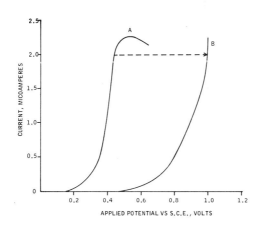

FIG. 6

Polarograms of H_2O_2 (A) and enzymatically
oxidized peroxide solutions (B) (ref. 97).

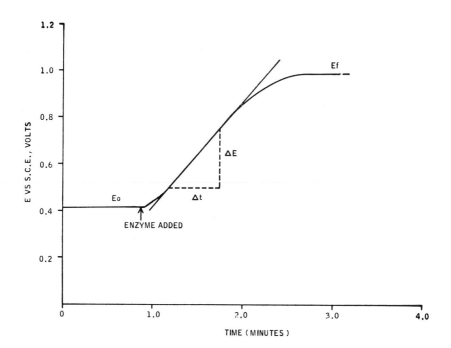

FIG. 7

Voltage-time curve for enzymatic
oxidation of peroxide (ref. 97).

observed for the oxidation products of peroxide (Figure
6B), is identical with the polarogram recorded for water.
From linear calibration plots of the change in potential with time, $\Delta E/\Delta t$, vs. concentration of catalase
the amount of enzyme present can be determined in the
range of 0.010 to 1.00 unit.

3. Glucose Oxidase

Glucose oxidase catalyzes the reaction:

$$\beta\text{-D-glucose} + H_2O + O_2 \longrightarrow \text{D-gluconic acid} + H_2O$$

The activity of the glucose oxidase can be measured
manometrically by noting the uptake of oxygen, electrochemically by measuring the potential or current change
in the enzymic system, or colorimetrically or fluorometrically using a peroxide coupled indicator reaction:

$$H_2O_2 + \text{Reduced dye} \xrightarrow{\text{Peroxidase}} \text{Oxidized dye} + 2\ H_2O$$

  (Colorless)   (Colored or fluorescent)

In the conventional manometric assay the oxygen
uptake is measured in a Warburg apparatus and is a
measure of the enzyme activity.[98]   Alternatively
glucose oxidase can be determined by measuring the
oxygen consumed with an oxygen electrode (Beckman model
777 or equivalent). The determination is possible with
catalase free oxidase or with a mixture of catalase and
oxidase.[99] Since peroxide derived from glucose
slowly decomposes and many samples of glucose oxidase
contain small amounts of catalase, excess catalase is
used to ensure more accurate results.[99] The polarographic oxygen electrode was used by Kadish and Hall[100]
and Makino and Koono[101], in a similar manner, for
measuring glucose oxidase and glucose in blood. Updike
and Hicks[102,103] used glucose oxidase immobilized in
an acrylamide gel and an oxygen electrode for monitor-

ing glucose.

Kadish, Litle and Sternberg[104] incorporated iodide, molybdate and ethanol with glucose oxidase for maximum sensitivity and reproducibility with the oxygen electrode:

$$\text{Glucose} + O_2 \xrightarrow{\text{Glucose Oxidase}} H_2O_2$$

$$2I^- + H_2O_2 + 2H^+ \xrightarrow{\text{ammonium molybdate}} I_2 + 2H_2O$$

Any catalase-catalyzed decomposition of peroxide leads to formation of acetaldehyde in the presence of alcohol, rather than release of oxygen.[105] This is an induced reaction resulting from radical attack on the ethanol.

Guilbault et al.[106] proposed a simple electrochemical method for the analysis of glucose oxidase. The apparatus described above for cholinesterase was used, with small constant current of 40 μa applied across two platinum thimble electrodes. Because the substrate, glucose, is electrochemically inactive in this system, diphenylamine sulfonic acid (DPASA) was used to establish a selected, well posed starting potential:

$$\text{Glucose} + \text{DPASA} \xrightarrow{\text{Glucose Oxidase}} H_2O_2 + \text{gluconic acid}$$
$$(E_o \sim 0.8v) \qquad\qquad\qquad (E_o \sim 0.4v)$$

Upon the enzymic catalyzed oxidation hydrogen peroxide is formed, which is electrochemically active ($E_o \sim 0.40v$) and the potential drops (Figure 8). By taking slopes of these depolarization curves, $\Delta E/\Delta t$, the glucose oxidase can be assayed over the range of concentrations 0.088 to 1.6 units per ml with a relative standard deviation of 1.5%.

Pardue and Malmstadt have developed automated electrochemical methods for the determination of glucose oxidase based on the oxidation of glucose to peroxide, followed

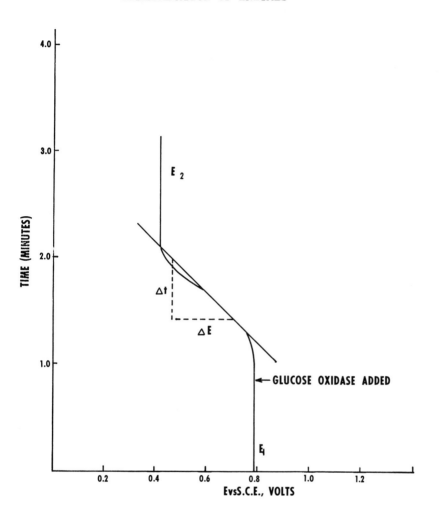

FIG. 8

Voltage-Time curve for the enzymic
oxidation of glucose by glucose oxidase.
(ref. 106)

by the oxidation of iodide to iodine in the presence
of molybdate as catalyst. The rate of production of
iodine is proportional to the rate of oxidation of
glucose, and is detected either potentiometrically[107-109] or amperometrically.[110,111] In either case,
automatic control equipment provides a direct read-
out of the time required for a predetermined amount
of iodine to be produced. The reciprocal of the time
interval is proportional to the glucose oxidase activity with a relative standard deviation of about 2%.
Blaedel and Olson developed a method for the assay of
glucose oxidase by an amperometric procedure similar
to the one described above, except that the peroxide
oxidizes ferrocyanide to ferricyanide, which is measured with a tubular platinum electrode.[112] Pardue
also extended the electrochemical techniques described to the assay of galactose oxidase.[113] The
peroxide produced again reacts with iodide to form
iodine, which is detected amperometrically. The reciprocal of the time interval required for a certain
current to be produced is proportional to the activity
of galactose oxidase.

Many colorimetric procedures have been described
for the analysis of glucose oxidase. The production
of a colored dye in a peroxide coupled reaction is
measured, and equated to glucose oxidase activity.
The most commonly used dye is o-dianisidine which is
oxidized to a highly colored product (λ_{max}=435-
460 mμ).[114] This dye is used in the conventional
Glucostat reagent (Worthington Biochem. Co.) and in
the Galactostat reagent using galactose oxidase.

Other dyes have been proposed as substitutes for
o-dianisidine: o-toluidine[115], 2-6-dichlorophen-
ol[116], o-tolidine[117,118] and a β-diketone (with
catalase)[119] to name but a few.

Guilbault et al.[120,121] have shown that homovanillic acid(XII)(4-hydroxy-3-methoxy phenylacetic acid) is an excellent fluorometric substrate for the determination of glucose oxidase. Upon oxidation homovanillic acid is converted to a highly fluorescent compound(XIII) which has a $\lambda_{ex}=315$ mμ; and a λ_{em} of 425 mμ:

$$\text{Glucose} + O_2 \xrightarrow{\text{glucose} \atop \text{oxidase}} H_2O_2$$

$$H_2O_2 + \text{Homovanillic Acid} \xrightarrow{\text{Peroxidase}}$$

$$\begin{array}{cc} & \text{Fluorescent Product} \\ \text{XII} & \text{XIII} \end{array}$$

The rate of production of oxidized homovanillic acid (XIII) is proportional to the concentration of glucose oxidase from 0.001 to 0.25 unit/ml. The homovanillic acid is completely stable in aqueous solution, and the same solution was used for 6 months with little apparent fluorescence formed in the reagent solution.

In attempting to develop more sensitive fluorometric methods for oxidative enzyme systems, Guilbault, Brignac and Juneau[122] tried 25 different substrates as possible replacements for homovanillic acid. Of these, p-hydroxyphenylacetic acid was judged to be the best substrate. It also is completely stable to auto-oxidation, and has advantages over homovanillic acid of low cost and a higher fluorescence coefficient (fluorescence/concentration in M). As little as 10^{-2} unit of galactose oxidase could be assayed.

4. Peroxidase

Peroxidase is an iron-porphyrin enzyme that reacts with hydrogen peroxide in two steps:

$$\text{Peroxide} \xrightarrow{\text{peroxidase}} \text{"oxidant"}$$

$$\text{"oxidant"} + DH_2 \longrightarrow \text{peroxidase} + H_2O + D$$

The first step is very selective; only hydrogen, methyl, and ethyl peroxides combine with the enzyme to form the

active enzyme-substrate "oxidant". The "oxidant" then
reacts with a reduced dye (DH) to yield oxidized dye,
which is highly colored or fluorescent.

The numerous peroxidase assay methods have been
reviewed by Maehly and Chance[123], and many of these
have been discussed above in the section on glucose
oxidase. In the most common colorimetric procedure,
the dye o-dianisidine is used as the hydrogen donor
(DH_2), and the colored oxidized dye is followed at 460 mμ.
One unit of peroxidase is that amount of enzyme decompos-
ing one micromole of peroxide per minute. Various other
dyes have been used in the colorimetric analysis of
peroxidase. In the analysis of peroxidase in cereal
and flour, the reduced 2,6-dichloroindophenol is used.
A blue color develops in the presence of peroxide.[124]
In milk guaicol[125], benzidine[126,127] and other
oxidizable substrates such as ascorbic acid,uric acid
cytochrome C and eugenol[128] have been used. Guilbault
and Kramer[129] described an improved substrate for the
rapid spectrophotometric assay of peroxidase. The compound, 4-methoxy-1-naphthol, is a colorless material
which is oxidized to an intensely blue colored compound
(ϵ = 1.8 x 10^4 at 620 mμ). From 0.01 to 0.170 unit
per ml of peroxidase and 0.0055 to 0.22 unit per ml of
glucose oxidase could be assayed. The oxidized material
was found to be a stable compound, whose color and
formation are independent of pH.

Likewise the potentiometric method outlined above
for catalase can be used for assay of peroxidase activity.[97] A small current of 2 μa applied across two
platinum electrodes immersed in a peroxide solution,
results in the establishment of a smooth initial potential of 0.40 v. Upon addition of peroxidase, the
potential increases, reaching the potential of oxygen
evolution from water (Figure 7). The slopes of these
curves, $\Delta E/\Delta t$, are proportional to the peroxidase

concentration over the range of activities 0.01 to 1.0 unit per ml. Since peroxidase requires a proton donor to function enzymatically, a donor was chosen whose oxidized and reduced forms are not electroactive at the potentials used (pyrogallol).

Fluorometrically, homovanillic acid (HVA) has been used as an indicator for measuring peroxidase activity as described above for glucose oxidase. The non-fluorescent HVA is oxidized to a highly fluorescent dimer, the rates of formation of the fluorescent product being proportional to the activity of peroxidase over the concentration range 10^{-3} to 2 units per ml.

Keston and Brandt[130] have described a fluorometric method for peroxidase, based on the oxidation of the non-fluorescent diacetyl dichlorofluorescin to the highly fluorescent dichlorofluorescein by hydrogen peroxide and peroxidase. The method is also applicable to other enzyme systems which produce hydrogen peroxide. Andreae[131] and Perschke and Broda[132] used scopoletin (6-methoxy-7-hydroxy-1,2-benzopyrene) as a substrate for peroxidase. The disappearance of the fluorescence of scopoletin was a measure of the peroxidase concentration.

5. Xanthine Oxidase

Xanthine oxidase catalyzes the oxidation of hypoxanthine to uric acid.

$$\text{Hypoxanthine} + O_2 \xrightarrow[\text{oxidase}]{\text{xanthine}} \text{xanthine} + H_2O_2$$

$$O_2 \Big\downarrow \begin{array}{l}\text{xanthine} \\ \text{oxidase}\end{array}$$

$$\text{uric acid} + H_2O_2$$

The enzyme is most commonly assayed by recording the rate of formation of uric acid from hypoxanthine spectrophotometrically at 290 mμ. A unit of activity is that forming one micromole of urate per minute at 25°C. [133]

Assay methods have also been described based on manometric measurement of oxygen uptake[134], or on reduction of methylene blue[135] or cytochrome c [136] in the absence of oxygen:

xanthine + Oxidized dye $\xrightarrow[\text{Oxidase}]{\text{Xanthine}}$

                         uric acid + Reduced Dye
        (Colored)   (Colorless)

Guilbault, Kramer and Cannon[137] proposed a simple electrochemical method for the analysis of xanthine oxidase. A small constant current of 3.8 µa is applied across two platinum electrodes, and the change in potential of the anode with time during the reaction is recorded. This small current establishes a steady reproducible potential with an electroinactive substrate, hypoxanthine, caused by oxidation of some component of the buffer solution, probably water. Upon addition of xanthine oxidase a decrease in potential is observed due to production of the more electroactive species, H_2O_2 and uric acid (Fig. 9). Calibration plots of $\Delta E/\Delta t$ vs. concentration of enzyme are linear over the range of 2×10^{-4} to 4×10^{-3} unit per ml. To test the reliability of the proposed method, a number of samples of xanthine oxidase from milk were tested, and the results obtained electrochemically agreed with those obtained colorimetrically (following the production of uric acid at 290 mµ) within 0.3% (Table 11).

Guilbault, Brignac and Zimmer[121] described a fluorometric method for the assay of 10^{-4} to 10^{-1} unit per ml of xanthine oxidase using homovanillic acid in a peroxidase coupled indicator reaction. The rate of production of the fluorescent oxidized HVA is proportional to the concentration of enzyme.

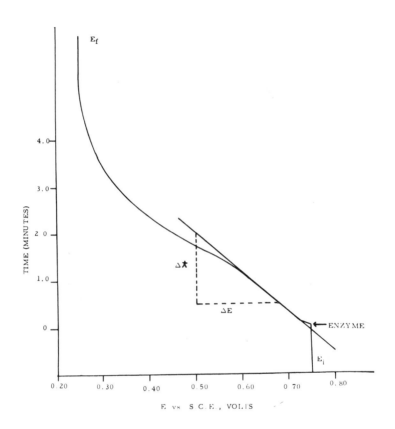

FIG. 9

Voltage-time curve for enzymatic oxidation
of hypoxanthine by xanthine oxidase.
(ref. 137)

TABLE 11

Analysis of Xanthine Oxidase Solutions

	Reported Activity units/ml.	Activity Found units/ml.	
		Electrochemically[a]	Colorimetrically
Worthington, lot No. X0755, milk	2.70	2.50	2.50
Worthington, lot No. X0771, milk	11.7	11.5	11.5
Calbiochem, B Grade, milk	8.0	7.50	7.52
Milk, purified [1]	----	0.501	0.505

[a]An average of 3 runs.

Lowry[138] has shown that 2-amino-4-hydroxy-pteridine is a good fluorogenic substrate for xanthine oxidase. Isoxanthopterin is formed, which is fluorescent. The substrate is non-fluorescent so that the rate of formation of fluorescence is proportional to enzyme concentration. The fluorophor can be assayed directly and continuously in the incubation mixture, and each sample acts as its own standard to correct for blank fluorescence or fluorescence losses.

A radiochemical method for assay of xanthine oxidase was proposed by Weinstein, Medes, and Litwack. [139] A radioactive amino acid substrate, $8\text{-}C^{14}$-xanthine was used as substrate for xanthine oxidase assay in blood serum and tissue.

D.   DEHYDROGENASE ENZYMES

1.   General

An important class of enzymes are the dehydrogenases, which in the presence of a hydrogen acceptor such as nicotinamide adenine dinucleotide (NAD)(DPN) or nicotinamide adenine dinucleotide phosphate (NADP)(TPN)

effect the dehydrogenation of hydroxy compounds. It
has been stated that almost every enzyme of biological
interest can be assayed with the aid of auxiliary
enzymes and the coenzymes NAD and NADPH.[140] Reduced
NADH has maximum absorbance at 340 mμ, while NAD has
little absorbance at this wavelength (Fig. 10). Hence,
a spectrophotometric measurement of NADH or NADPH

Substrate + NAD or NADP $\xrightarrow{\text{dehydrogenase}}$

$$\text{NADH + oxidized substrate}$$
$$\text{or}$$
$$\text{(NADPH)}$$

indicates the progress of the enzymic reaction.

Likewise all NAD and NADP-dependent dehydrogenases
can be monitored fluorometrically since NADH has a
high fluorescence (λ_{ex}=340 mμ; λ_{em}=450 mμ)(Fig. 11).
Lowry et al.[140] have described procedures for using
the fluorescence of NADH and NADPH for the assay of
enzymes and have discussed the effect of solvents, pH,
and trace metals on the fluorescence. The greater sensitivity of fluorescence permits the use of smaller
samples and smaller amounts of expensive substrates
and cofactors. At least 2 orders of magnitude increase in sensitivity is achieved over colorimetric
methods. The appearance of fluorescence or the loss
of fluorescence may be measured directly in the reaction mixture, either kinetically or after a predetermined period, in a manner completely analogous to the
measurements of absorbance at 340 mμ. The upper limit
of concentration in the final solution to be measured,
for which the fluorescence of NADH or NADPH would be
strictly linear with concentration, is of the order of
2.5 μg/ml. Ideally, conditions in kinetic methods
should be arranged such that the concentration of NADH
or NADPH does not exceed this concentration.

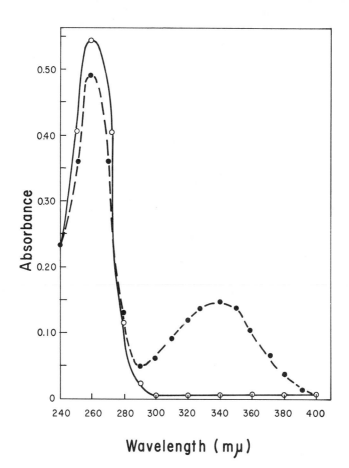

FIG. 10

Absorption curve of NAD (DPN)(o----o)
and NADH (DPNH)(●----●).

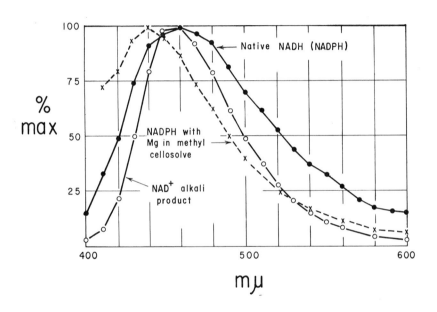

FIG. 11

Fluorescence spectra of pyridine nucleotides.
Values are plotted as per cent of the maximum
fluorescence. On an absolute scale NADH
fluorescence would have to be reduced by a
factor of about 8 (ref. 140).

Kaplan et al.[141] have found that NAD and NADP are converted to highly fluorescent products when heated with alkali, and Lowry et al.[140] stabilized the fluorophor and have reported its fluorescence characteristics (Fig. 11). Huff and Perlzweig[142] have shown that NAD and NADP condense with acetone in alkaline solution to form highly fluorescent products.

Thus the activity of a dehydrogenase can be determined fluorometrically, either by monitoring the fluorescence of NADH or NADPH directly, or by a measurement of NAD or NADP after treatment with alkali or acetone.

A simple, rapid fluorometric method was described by Guilbault and Kramer[143,144] for measuring the activity of dehydrogenases. The method is based on the conversion of the non-fluorescent material resazurin(XIV) to the highly fluorescent resorufin(XV) in conjunction with the NAD-NADH or NADP-NADPH system. As little as 10^{-4} unit per ml of the enzymes lactic dehydrogenase, alcohol dehydrogenase, malic dehydro-

$$\text{Substrate + NAD} \xrightarrow{\text{dehydrogenase}} \text{oxidized substrate + NADH}$$
(NADP)   (NADPH)

NADH +

XIV
Resazurin
(Non-Fluorescent)

$\xrightarrow{\text{Phenazine Methyl Sulfate}}$

XV
Resorufin
(λ_{ex} = 540 mμ; λ_{em} = 580 mμ)

+ NAD

genase, glutamate dehydrogenase, glucose 6-phosphate dehydrogenase, L-α-glycerophosphate dehydrogenase and glycerol dehydrogenase could be determined with standard deviations of less than one percent (Table 12). Likewise NADH (10^{-7} to 10^{-5}M) may be determined with

TABLE 12
Determination of Various Dehydrogenases

Dehydrogenase	Amount Added, Units/ml.	Standard Deviation, %
Lactic	0.0003-0.100	1.1
Malic	0.00105-0.510	1.4
Alcohol	0.0003-0.151	0.8
Glutamate	0.0001-0.033	0.9
Glucose-6-phosphate	0.002-0.340	1.1
Glycerol phosphate	0.010-1.10	1.1
Glycerol	0.005-0.505	0.9

a standard deviation of about 0.5%.

Because of the intense fluorescence of resorufin (as little as $10^{-9}M$ can be detected), an increase in sensitivity of one to two orders of magnitude over the NADH fluorescence method is achieved. Resazurin has been successfully used as a substrate for dehydrogenase in automated systems developed by Technicon.[145] The resorufin formed is continually monitored.

2. Lactate Dehydrogenase

Measurement of serum lactate dehydrogenase (LDH) is useful in the diagnosis of several potentially fatal chemical entities. Serum LDH has been shown to rise in myocardial infraction[146,147], pulmonary infraction[148], hepatitis[149], congestive heart failure[150], leukemia[151] and certain other malignancies.[152]

Serum lactate dehydrogenase catalyzes the conversion of lactate to pyruvate:

$$CH_3-CHOH-COOH + NAD \xrightarrow{\text{LDH}} CH_3-CO-COOH + NADH + H^+$$

(Fluorescent)

The reaction is quantitative from left to right at a pH
of 9, but the reverse reaction is effected at a pH of 7.
At a pH of 9 the non-fluorescent NAD is reduced to NADH
which is fluorescent. The rate of formation of NADH is
proportional to the concentration of LDH.[153,154] The
diagnostic specificity of a serum LDH assay is improved
by measuring the individual LDH isoenzymes. A complete
discussion of the fluorometric determination of all five
LDH isoenzymes separated by electrophoresis is available[155,156]

Hicks and Updike[157] have conducted a thorough study
of LDH determination in urine and found that a non-enzymic
reducing substance is encountered when measuring urine
LDH by the forward reaction, lactate to pyruvate. Because NADH (DPNH) is used in the reverse reaction, this
positive interference is not encountered. Brooks and
Olken[158] have described an automated fluorometric
method for measuring LDH. After a 5 minute incubation,
the fluorescence of NADH was measured.

In a similar manner the activity of lactate dehydrogenase can be monitored fluorometrically using the
resazurin-resorufin indicator reaction.[144] As little
as 10^{-4} unit of LDH is determinable.

3. Glucose-6-Phosphate Dehydrogenase

Glucose-6-phosphate dehydrogenase (G-6-P-DH) catalyzes
the reaction:

$$\text{Glucose-6-Phosphate} + \text{NADP} \xrightarrow{\text{G6P-DH}}$$
$$\text{6-phosphogluconate} + \text{NADPH} + \text{H}^+$$

G6P-DH has been demonstrated in almost all animal tissues
and in micro-organisms. The rate of formation of NADPH
is a measure of the enzyme activity and it can be
followed by means of the increase in absorption at 340
or 366 mμ[159] or fluorometrically.

Guilbault and Kramer[144] have described a fluoro-

metric method for G6P-DH using the resazurin-resorufin
indicator reaction. As little as 10^{-4} unit of enzyme
is determinable.

4. Glutamate Dehydrogenase

Glutamate dehydrogenase (GDH) catalyzes the reaction:

$$\alpha\text{-Oxoglutarate} + NH_4^+ + NADH \xrightleftharpoons{GDH} L\text{-glutamate} + NAD + H_2O$$

The equilibrium lies in favor of the amino acid formation.

Oxidation of NADH is directly proportional to the
reduction of substrate and can be followed by a decrease
in absorbance at 340 or 366 mμ.[160]

Alternatively the rate of production of the fluorescence of resorufin is proportional to the concentration
of GDH[144]:

$$L\text{-glutamate} + NAD + H_2O \xrightarrow{GDH} \alpha\text{-Oxoglutarate} + NH_4^+ + NADH$$

$$NADH + Resazurin \xrightarrow{\text{Phenazine Methyl Sulfate}} Resorufin$$

    (Non-Fluorescent)   (Fluorescent)

The equilibrium of the first reaction is displaced in
favor of formation of NADH and α-oxoglutarate by (a)
coupling with the resazurin-resorufin indicator reaction and (b) by adding hydrazine to complex the α-oxo-
glutarate.

5. α-Hydroxy Butyrate Dehydrogenase

A number of recent papers have indicated that α-
hydroxy butyrate hydrogenase (HBD) is the single best
enzyme test for confirmation of myocardial necrosis.[161]
An HBD test is more definitive than total LDH, SGOT or
SGPT. Since HBD catalyzes the reduction of α-ketobutyric
acid to α-hydroxy butyric with simultaneous oxidation of
NADH to NAD, the reaction can be followed kinetically
by the fluorescence of NADH.[161] The assay is rapid
(1 minute) and linear from 0-2,000 International

Units:

$$\text{α-ketobutyrate + NADH} \xrightleftharpoons{\text{BDH}} \text{α-hydroxybutyric + NAD}$$

(Fluorescent)

Likewise the reaction can be monitored spectrophoto-metrically by the decrease in absorbance at 340 or 366 mμ.

6. Alcohol and Glycerol Dehydrogenases

Alcohol dehydrogenase catalyzes the oxidation of ethanol to acetaldehyde[162]; glycerol dehydrogenase that of glycerol to dihydroxyacetone.[163] Both enzymes can be assayed either fluorometrically[164] or spectrophotometrically by recording the rate of production of NADH with time.[162,163]

$$\text{Ethanol + NAD} \xrightleftharpoons{\text{ADH}} \text{RCHO + NADH}$$

$$\text{Glycerol + NAD} \xrightleftharpoons{\text{GDH}} \text{dihydroxyacetone + NADH}$$

Likewise both enzymes can be measured fluorometrically with added sensitivity using the resazurin-resorufin reaction.[144] The rate of production of the highly fluorescent resorufin is proportional to the concentra-

$$\text{NADH + Resazurin} \xrightleftharpoons[\text{Sulfate}]{\text{Phenazine Methyl}} \text{Resorufin}$$

(Non-Fluorescent)   (Fluorescent)

tion of alcohol dehydrogenase (ADH) or glycerol dehydrogenase (GDH). As little as 10^{-4} unit of enzyme is determinable.

7. Malate Dehydrogenase

Malate dehydrogenase (MDH) catalyzes the following reaction:

$$\text{L-Malate + NAD} \xrightleftharpoons{\text{MDH}} \text{oxaloacetate + NADH.}$$

Since the equilibrium of this reaction lies far to the left, the measurements of MDH activity are made with oxaloacetate as substrate and NADH as coenzyme. The decrease in absorbance at 340 or 366 mμ is measured.[165]

Guilbault and Kramer[144] have proposed a fluorometric assay of as little as 10^{-4} unit of MDH. The resazurin-resorufin indicator reaction at high pH is used to displace the equilibrium in favor of resorufin formation.

$$NADH + Resazurin \xrightarrow[Sulfate]{Phenazine\ Methyl} Resorufin + NAD$$
$$(Fluorescent)$$

The rate of increase of fluorescence with time, $\Delta F/min$, is proportional to the MDH concentration.

E. KINASES AND TRANSAMINASES

Kinases are enzymes which catalyze the transfer of a phosphate group, usually in the presence of a coenzyme such as adenosine triphosphate (ATP), adenosine diphosphate (ADP), etc. Acetokinase for example effects the phosphorylation of acetate:

$$ATP + acetate \xrightarrow{acetokinase} ADP + acetyl\ phosphate$$

Glycerokinase catalyzes the phosphorylation of glycerol, creatine kinase that of creatine, etc.

$$Glycerol + ATP \xrightarrow{Glycerokinase} L-\alpha\text{-glycerophosphate} + ADP$$

Most kinases can be assayed by coupling with a dehydrogenase system. The production or disappearance of NADPH or NADH fluorometrically or colorimetrically then serves as a measure of the concentration of the kinase.

Glycerokinase, for example, can be determined by coupling with the glycerophosphate dehydrogenase system.[166] The L-α-glycerophosphate formed is oxidized to dihydroxy acetone phosphate with concomitant formation of NADH:

L-α-glycerophosphate + NAD $\xrightarrow{\text{dehydrogenase}}$

$$\text{dihydroxy acetone phosphate +}$$
$$\text{NADH + H}^+$$

Correspondingly a second kinase system plus a dehydrogenase system can be linked together for the assay of a kinase. Sherwin, Siber and Elhilai[167] for example, developed a fluorometric technique for creatine kinase using a hexokinase-glucose-6-phosphate dehydrogenase coupled indicator reaction:

Creatine phosphate + ADP $\xrightarrow{\text{kinase}}$ creatine + ATP

Glucose + ATP $\xrightarrow{\text{hexokinase}}$ glucose-6-phosphate + ADP

Glucose-6-phosphate + NADP $\xrightarrow{\text{dehydrogenase}}$

$$\text{NADPH + 6-phosphogluconate}$$

The rate of production of the fluorescence of NADPH is a measure of the creatine kinase activity.

Methods have likewise been proposed for the assay of kinases using other systems. Sax and Moore for example, assayed creatine kinase by monitoring the creatine liberated with a fluorometric ninhydrin indicator reaction. A fluorophor is formed in alkaline solution.[168] This same type of assay was used by Conn and Anido for the determination of creatine kinase.[169]

Transaminases catalyze transamination (transfer of an amino group), a general metabolic reaction for the synthesis and degradation of L-amino acids.

$$\begin{matrix} R \\ H \\ R \end{matrix}\!\!\diagdown\!\!C\text{-NH}_3^+ + \begin{matrix} R' \\ R' \end{matrix}\!\!\diagdown\!\!C\text{=O} \rightleftharpoons \begin{matrix} R \\ R \end{matrix}\!\!\diagdown\!\!C\text{=O} + H\begin{matrix} R' \\ R' \end{matrix}\!\!\diagdown\!\!C\text{-NH}_3^+$$

General assay methods for transaminases have been reviewed in detail by Aspen and Meister.[170]

Two of the most common transaminases are glutamate-oxaloacetate and glutamate pyruvate. Glutamate-oxaloacetate transaminase (GOT) catalyzes the reaction:

L-glutamate + oxaloacetate $\xrightleftharpoons{\text{GOT}}$ L-aspartate + α-oxoglutarate

The activity of the enzyme can be measured by coupling this reaction with the malic dehydrogenase system.

Oxaloacetate + NADH + H$^+$ $\xrightarrow{\text{Malic}}_{\text{Dehydrogenase}}$ malate + NAD

The rate of decrease of NADH (spectrophotometrically at 340 or 366 mμ or fluorometrically) is a measure of the concentration of transaminase.[171]

Graham and Aprison,[172] and Passen and Gennaro[173] monitored glutamate-oxaloacetate transaminase by measuring the decrease of NADH fluorometrically. Likewise the increase in NAD could be measured fluorometrically.

The GOT activity can also be measured colorimetrically. The oxaloacetate is quantitatively converted to pyruvate with anilinium citrate as catalyst.[174] The pyruvate is then determined as the 2,4-dinitrophenylhydrazone (λ_{max} = 450 mμ). Or more simply, the 2,4-dinitrophenylhydrazone of the reaction product, oxaloacetate, can be measured directly in alkaline solution[175] (λ_{max} = 530 or 546 mμ).

Glutamate-pyruvate transaminase (GPT) catalyzes the reaction:

L-glutamate + pyruvate $\xrightleftharpoons{}$ L-alanine + α-oxoglutarate

The most commonly accepted method for assay of GPT is based on a measurement of pyruvate formed from alanine and α-oxoglutarate, either with a lactic dehydrogenase coupled enzyme system, or colorimetrically with 2,4-dinitrophenyldrazine.

In the first method the pyruvate formed is converted to lactate with concommitant formation of NAD from NADH.[176,177]

$$\text{Pyruvate + NADH + H}^+ \rightleftharpoons \text{lactate + NAD}$$

The rate of decrease in the absorbance of NADH at 340 mμ, or the increase in the fluorescence of NAD[178,179] is measured and equated to the amount of transaminase present.

Like the GOT activity, GPT activity can be measured colorimetrically by reacting pyruvate with 2,4-dinitrophenylhydrazine.[180,181] Measurements are not made at the λ_{max} of pyruvate hydrazone but at 500-550 mμ in order to minimize color formation from the α-oxoglutarate hydrazone.

Pitts, Quick and Robins[182] coupled the NAD linked succinic semialdehyde dehydrogenase reaction with a transaminase reaction to measure γ-aminobutyric-α-oxoglutaric transaminase. The NADH produced was measured fluorometrically.

F.   OTHER ENZYMES

In this chapter attempts were made to give the reader a survey of methods available for many of the more important enzymes. It is realized that this listing is not a comprehensive one. Details on methods for other enzymes can be found in reviews by Guilbault[183,184], a book by Bergmeyer[185] or in handbooks of procedures distributed by biochemical companies selling enzyme products (Worthington).

1. P. Bernfeld, Methods in Enzymology (S. P. Colowick and N. O. Kaplan, etd.), Academic Press, New York, 1955, p. 149.

2. H. V. Street, Clin. Chim. Acta 3, 501 (1958).

3. H. V. Street, Clin. Chim. Acta 1, 256 (1956).

4. R. Ammon and G. Voss, Pflugers Arch ges Physiol. Menschen Tiere 235, 393 (1935).

5. D. N. Kramer, R. M. Gamson, Anal. Chem. 30, 251 (1958).

6. J. de la Huerga, Ch. Yesinick and H. Popper, Amer. J. Clin. Pathol. 22, 1126 (1952).

7. G. Ellman, Biochem Pharmacol. 7, 88 (1961).

8. D. N. Kramer, P. L. Cannon and G. G. Guilbault, Anal. Chem. 34, 842 (1962).

9. G. G. Guilbault and D. N. Kramer, Anal. Biochem 5, 208 (1963).

10. G. G. Guilbault, D. N. Kramer and P. L. Goldberg, J. Phys. Chem. 67, 1747 (1963).

11. G. G. Guilbault and D. N. Kramer, Anal. Chem. 36, 1662 (1964).

12. G. G. Guilbault and D. N. Kramer, Anal. Chem. 37, 120 (1965).

13. G. G. Guilbault and D. N. Kramer, Anal. Chem. 38, 1675 (1966).

14. G. G. Guilbault and M. H. Sadar, R. Glazer and C. Skou, Anal. Letters, 1, 367 (1968).

15. F. P. Winteringham and R. W. Disney, Nature 195, 1303 (1962).

16. D. J. Reed, K. Goto, C. H. Wang, Anal. Biochem. 16, 59 (1966).

17. L. T. Potter, J. Pharm. Exp. Ther. 156, 500 (1967).

18. M. Mandels and E. T. Reese, Industrial Microbiol. 5, 5 (1964).

19. B. Norkraus, Physiol. Plant 3, 75 (1950).

20. S. Meyers, B. Prindle and E. Reynolds, Tappi 43, 534 (1960).

21. G. G. Guilbault and A. Heyn, Anal. Letters 1, 163 (1967).

22. P. Desnuelle, in The Enzymes, Vol. 4, (P. D. Boyer, H. Lardy and K. Myrback, eds.), Academic Press, Inc., New York, 1960, p. 93.

23. G. R. Schonbaum, B. Zerner and M. L. Bender, J. Biol. Chem. 236, 2930 (1961).

24. B. F. Erlanger and F. Edel, Biochem. 3, 346 (1964).

25. C. L. Martin, J. Golubow, and A. E. Axelhod, J. Biol. Chem. 234, 294 (1959).

26. H. F. Bundy, Anal. Biochem. 3, 431 (1962).

27. G. W. Schwert and Y. Takenaka, Biochim. Biophys Acta 26, 570 (1955).

28. B. C. W. Humme. Can. J. Biochem. Physiol. 37, 1393 (1959).

29. B. H. Bielski and S. Freed, Anal. Biochem. 7, 192 (1964).

30. G. G. Guilbault and D. N. Kramer, Anal. Chem. 36, 409 (1964).

31. J. Larner, The Enzymes, Vol. 4, 2nd ed., (P.D. Boyer, H. Lardy and K. Myrback, eds.), Academic Press, Inc., New York, 1960, p. 369.

32. S. Veibel, The Enzymes, Vol. 1, Part 1 (J. B. Summer and K. Myrback, eds.), chapter 16, Academic Press, Inc., New York, (1950).

33. N. N. Nelson, J. Biol. Chem. 153, 375 (1944).

34. E. Hofmann and G. Hofmann, Biochem. Z. 324, 397 (1953).

35. G. G. Guilbault and D. N. Kramer, Anal. Biochem. 18, 313 (1967).

36. D. Robinson, Biochem. J. 63, 39 (1956).

37. J. W. Woolen and P. G. Walker, Clin. Chim. Acta 12, 647 (1965).

38. B. Rotman, J. A. Zderic, M. Edelstein, Proc. Natl. Acad. Sci. 50, 1 (1963).

39. W. H. Fishman, B. Springer and R. Brunetti, J. Biol. Chem. 173, 449 (1948),

40. J. W. Woolen and P. Turner, Clin. Chim. Acta 12, 659 (1965).

41. Ibid, p. 671.

42. J. A. Mead, J. N. Smith and R. T. Williams, Biochem. J. 61, 569 (1955).

43. M. Veritz, R. Caper and W. Brown, Arch. Biochem. Biophys. 106, 386 (1964).

44. L. J. Greenberg, Anal. Biochem. 14 265 (1966).

45. P. Hoffman, K. Meyer and A. Linker, J. Biol. Chem. 219, 653 (1956).

46. F. Duran-Reynals, Ann. N. Y. Acad. Sci. 52, 943 (1950).

47. N. Ferrante, J. Biol. Chem. 220, 303 (1956).

48. S. Tolksdorf, M. McCready, D. R. McCullagh and E. Schwenk, J. Lab. Clin. Med. 34, 74 (1949).

49. D. McClean, Biochem. J. 37, 169 (1943).

50. G. I. Swyer and C. W. Emmens, Ibid, 41, 29 (1947).

51. M. Rapport, K. Meyer and A. Linker, J. Biol. Chem. 186, 615 (1950).

52. G. G. Guilbault, D. N. Kramer and E. Hackley, Anal. Biochem. 18, 241 (1967).

53. G. G. Guilbault and D. N. Kramer, Anal. Biochem 14, 28 (1966).

54. S. P. Kramer, M. Bartalos, J. N. Karpa, J. M. Midel, A. Chang, A. M. Seligman, J. Surgical Research 4, 23 (1964).

55. H. A. Ravin and A. M. Seligman, Arch. Biochem. Biophys. 42, 337 (1953).

56. D. N. Kramer and G. G. Guilbault, Anal. Chem. 35, 588 (1963).

57. G. G. Guilbault and D. N. Kramer, Anal. Chem. 36, 409 (1964).

58. J. D. Sapira and A. P. Shapiro, Am. Fed. Clin. Res., Section Meeting, (1964).

59. R. E. Phillips and F. R. Elevitch, Fluorometric Techniques in Clinical Pathology and Their Interpretation, in Progress in Clinical Pathology. (M. Stefanini, ed.), Grune and Stratton, New York, 1966, pp. 118-122.

60. T. J. Jacks and H. W. Kircher, Anal. Biochem. 21, 270 (1967).

61. G. G. Guilbault and M. H. Sadar, Anal. Letters 1, 551 (1968).

62. C. Huggins and P. Talalay, J. Biol. Chem. 159, 399 (1945).

63. K. Linhard and K. Walter, Z. Physiol. Chem. 289, 245 (1952).

64. J. H. Wilkinson and A. V. Vodden, Clin. Chem. 12, 701 (1966).

65. C. M. Coleman, Clin. Chim. Acta 13, 401 (1966).

66. O. A. Bessey, O. H. Lowry and M. J. Brock, J. Biol. Chem. 164, 321 (1946).

67. M. A. Andersch and A. J. Szcypinski, Amer. J. Clin. Pathol. 17, 571 (1947).

68. H. Neuman, M. Van Vreedendaal, Clin. Chem. Acta 17, 183 (1967).

69. T. V. Hausman, R. Helger, W. Rick and W. Gross, Clin. Chim. Acta 15, 241 (1967).

70. H. Scharer, J. Dairy Sci. 21, 21 (1938).

71. G. Schwartz and O. Fischer, Milchwiss 3, 41 (1948).

72. D. W. Moss, Clin. Chim. Acta 5, 283 (1960).

73. L. J. Greenberg, Biochem. Biophys. Res. Commun. 9, 430 (1962).

74. T. Takeuchi and S. Nogami, Acta Pathol. Japan 4, 277 (1954).

75. F. R. Elevitch, S. Aronson, T. V. Feichtmeir and M. L. Enterline, 113th Annual Meeting, Am. Med. Assoc., San Francisco, June 22-25, 1964; Tech. Bull. Reg. Med. Tech. 36, 282 (1966).

76. D. B. Land and E. Jackim, Anal. Biochem. 16, 481 (1966).

77. H. N. Fernley and P. G. Walker, Biochem. J. 97, 95 (1965).

78. G. G. Guilbault, S. H. Sadar, R. Glazer and J. Haynes, Anal. Letters 1, 333 (1968).

79. J. Summer, J. Biol. Chem. 69, 435 (1926).

80. G. Gorin, E. Fuchs, L. G. Butler, S. L. Chopra and R. T. Hersh, Biochem. 1, 911 (1962).

81. A. Shatalova and G. I. Meerov, Biokhimiya 28, 384 (1963).

82. S. A. Katz, Anal. Chem. 36, 2500 (1964).

83. S. Katz and J. Cowans, Biochim. Biophys. Acta. 107, 605 (1965).

84. W. C. Purdy, G. D. Christian, E. C. Knoblock, Presented at the Northeast Section, American Association of Clinical Chemists, 16th National Meeting, Boston, Mass., August 17-20, 1964.

85. A. B. Crowther and B. S. Large, Analyst 81, 64 (1956).

86. D. Wellner and A. Meister, J. Biol. Chem. 235, 2013 (1960).

87. G. G. Guilbault and J. Montalvo, Anal. Chem., in press.

88. H. V. Malmstadt and T. P. Hadjiioannou, Anal. Chem. 35, 14 (1963).

89. G. G. Guilbault and J. Hieserman,, Anal. Biochem., 26, 1 (1968).

90. A. C. Maehly and B. Chance, Methods of Biochemical Analysis, Vol. 1, (D. Glick, ed.), Interscience, New York, 1954, p.357.

91. B. Chance and A. C. Maehly, Methods in Enzymology, Vol. II, (S. P. Colowick and N. O. Kaplan, eds.), Academic Press, New York, 1955, p. 764.

92. D. Appleman, Anal. Chem. 23, 1627 (1951).

93. R. F. Beers and I. W. Sizer, Science 117, 710 (1953).

94. O. Lobeck, Milchwirtsch Zbl. 6, 316 (1910).

95. R. F. Beers and I. W. Sizer, J. Biol. Chem. 195, 133 (1952).

96. K. G. Stern, Z. Physiol. Chem. 204, 259 (1932).

97. G. G. Guilbault, Anal. Biochem. 14, 61 (1966).

98. D. Keilin and E. F. Hartree, Biochem. J. 42, 230 (1948).

99. T. Kajihara and B. Hagihara, Rinsho Byori. 14 (4), 322 (1966).

100. A. H. Kadish and D. A. Hall, Clin. Chem. 9, 869 (1965).

101. Y. Makino and K. Koono, Rinsho Byori 15, 391 (1967).

102. S. J. Updike and G. P. Hicks, Nature 214, 986 (1967).

103. Ibid., Science 158, 270 (1967).

104. A. Kadish, R. Litle and J. C. Sternberg, Clin. Chem. 14, 116 (1968).

105. D. Keilin and E. Hartree, Biochem. J. 42, 230 (1948).

106. G. G. Guilbault, B. C. Tyson, D. N. Kramer and P. L. Cannon, Anal. Chem. 35, 582 (1963).

107. H. V. Malmstadt and H. L. Pardue, Anal. Chem. 33, 1040 (1961).

108. Ibid., Clin. Chem. 8, 606 (1962).

109. H. Pardue, R. Simon and H. Malmstadt, Anal. Chem. 36, 735 (1964).

110. H. Pardue, Anal. Chem. 35, 1240 (1963).

111. H. Pardue and R. Simon, Anal. Biochem. 9, 204 (1964).

112. W. J. Blaedel and C. Olson, Anal. Chem. 36, 343 (1964).

113. H. Pardue and C. Frings, J. Electroanal. Chem. 7, 398 (1964).

114. A. St. G. Hugget and D. A. Nizon, Biochem. J. 66, 12P (1957).

115. L. L. Salomon and J. E. Johnson, Anal. Chem. 31, 453 (1959).

116. L. A. Dobrick, J. Biol. Chem. 231, 403 (1958).

117. R. Thompson, Clin. Chim. Acta, 13, 133 (1966).

118. E. Kawerau, Z. Klin. Chem. 4, 224 (1966).

119. Fyowa Fermentation Industry, French Patent 1,410,747, (1967).

120. G. G. Guilbault, D. N. Kramer and E. Hackley, Anal. Chem. 39, 271 (1967).

121. G. G. Guilbault, P. Brignac and M. Zimmer, Anal. Chem. 40, 190 (1968).

122. G. G. Guilbault, P. Brignac, and M. Juneau, Anal. Chem. 40, 1256 (1968).

123. A. C. Maehly and B. Chance, Methods of Biochemical Analysis, Vol. I (D. Glick, ed.), Interscience, New York, 1954, p. 357.

124. F. Herrlinger and F. Kiesmeier, Biochem. Z. 317, 1 (1944).

125. B. Chance and A. C. Maehly, Methods in Enzymology, (S. Colowick and N. Kaplan eds.), Vol. II, Academic Press, New York, 1955, p. 764.

126. S. Rothenfusser, Z. Unters Lebensm 16, 74 (1908).

127. F. Bengen, Z. Unter Lebensm 66, 126 (1933).

128. B. Chance and A. C. Maehly, Biochemist Handbook, C. Long, ed., Van Nostrand, Princeton, 1961, p.384.

129. G. G. Guilbault and D. N. Kramer, Anal. Chem. 36, 2494 (1964).

130. A. S. Keston and R. Brandt, Anal. Biochem. 11, 1 (1965).

131. W. A. Andreae, Nature, 175, 859 (1955).

132. H. Perschke and E. Broda, Nature 170, 257 (1961).

133. H. M. Kalckar, J. Biol. Chem. 167, 429 (1947).

134. E. G. Gall, J. Biol. Chem. 128, 51 (1939).

135. M. Dixon and S. Thurlow, Biochem. J. 18, 976 (1924).

136. B. L. Horecker and L. A. Heppel, J. Biol. Chem. 178, 683 (1949).

137. G. G. Guilbault, D. N. Kramer and P. Cannon, Anal. Chem. 36, 606 (1964).

138. O. H. Lowry, O. A. Bessey and E. J. Crawford, J. Biol. Chem. 180, 399 (1949).

139. A. Weinstein, G. Medes and G. Litwack, Anal. Biochem. 21, 86 (1967).

140. O. H. Lowry, N. R. Roberts and J. I. Kapphahn, J. Biol. Chem. 224, 1047 (1957).

141. N. O. Kaplan, S. P. Colowick and C. C. Barnes, J. Biol. Chem. 191, 461 (1951).

142. J. W. Huff and W. A. Perlzweig, J. Biol. Chem. 167, 157 (1947).

143. G. G. Guilbault and D. N. Kramer, Anal. Chem. 36, 2497 (1964).

144. Ibid., 37, 1219 (1965).

145. Private Communication, Technicon Company, New York.

146. W. Wacker, D. Ulmer and B. L. Vallee, New Eng. J. Med. 225 , 449 (1956).

147. F. Wroblewski and J. S. La Due, Proc. Soc. Exp. Biol. Med. 90, 210 (1955).

148. W. Wacker and P. Snodgrass, J. Am. Med. Assoc. 174, 2142 (1960).

149. T. Laursen, Scand. J. Clin. Lab. Invest. 11, 134 (1959).

150. H. Refsum, Chim. Sci. 25, 369 (1963).

151. H. Bierman, B. Hill, L. Reinhardt and E. Emory, Cancer Res. 17, 660 (1957).

152. F. Wroblewski, Cancer 12, 27 (1959).

153. S. Ochoa, A. H. Mahler, and A. Kornberg, J. Biol. Chem. 174, 979 (1948).

154. F. R. Elevitch and R. E. Phillips, Fluorometric Method for LDH in Serum, G. K. Turner Associates, Palo Alto, California, (1966).

155. F. R. Elevitch, Thin Gel Electrophoresis, G. K. Turner Associates, Palo Alto, California, (1964).

156. J. Bergerman, Clin. Chem. 12, 797 (1966).

157. G. P. Hicks and S. J. Updike, Anal. Biochem. 10, 290 (1965).

158. L. Brooks and H. G. Olken, Clin. Chem. 11, 748 (1965).

159. O. Warburg, W. Christian and A. Griese, Biochem. Z. 282, 157 (1935).

160. C. Freiden, J. Biol. Chem. 234, 809 (1959).

161. P. A. Benson and W. H. Benedict, Am. J. Clin. Path. 45, 760 (1966).

162. B. L. Vallee and F. L. Hick, Proc. Nat. Acad. Sci. 41, 327 (1955).

163. R. M. Burton, Methods in Enzymology, Vol. 1 (S. P. Colowick and N. O. Kaplan, eds.), Academic Press, New York, 1955, p. 397.

164. G. R. Morrison and F. E. Brock, J. Lab. Clin. Med. 70, 116 (1967).

165. A. M. Hehler, A. Kornberg, S. Grisola and S. Ochoa, J. Biol. Chem. 174, 961 (1948).

166. O. Wieland, Biochem. Z. 329, 313 (1957).

167. A. L. Sherwin, G. R. Siber and M. M. Elhilai, Clin. Chem. Acta 17, 245 (1967).

168. S. M. Sax and J. Moore, J. Clin. Chem. 11, 951 (1965).

169. R. B. Conn and V. Anido, Am. J. Clin. Pathol. 42, 177 (1966).

170. A. J. Aspen and A. Meister, Methods of Biochemical Analysis, Vol. 6, Interscience, New York, 1958.

171. A. Karmen, J. Clin. Invest. 34, 131 (1955).

172. L. T. Graham and M. H. Aprison, Anal. Biochem. 15, 487 (1966).

173. S. Passen and W. Gennaro, Am. J. Clin. Pathol. 46, 69 (1966).

174. N. E. Tonhazy, N. G. White and W. W. Umbreit, Arch. Biochem. Biophysics 28, 36 (1950).

175. A. P. Hansen, Nordisk Med. 61, 799(1959).

176. K. S. Henley and H. M. Pollard, J. Lab. Clin. Med. 46, 785 (1955).

177. F. Wroblewski and J. S. La Due, Proc. Soc. Expo. Biol. Med. 91, 569 (1956).

178. T. Laursen and P. F. Hansen, Scand. J. Clin. Lab. Invest. 10, 53 (1958).

179. T. Laursen and G. Espersen, Ibid., 11, 61 (1959).

180. S. Reitman and S. Frankel, Amer. J. Clin. Pathol. 28, 56 (1957).

181. F. Wroblewski and P. Cabaud, Amer. J. Clin. Pathol. 27, 235 (1957).

182. F. N. Pitts, C. Quick and E. Robins, J. Neurochem. 12, 93 (1965).

183. G. G. Guilbault, Anal. Chem. $\underline{38}$, 529R (1966).

184. G. G. Guilbault, Anal. Chem. $\underline{40}$, 459R (1968).

185. H. U. Bergmeyer, Methods of Enzymatic Analysis,
     Academic Press, New York and London, 1965.

CHAPTER 4

DETERMINATION OF SUBSTRATES

A. GENERAL

At a fixed enzyme concentration, the initial rate
of enzymatic reaction increases with increasing substrate concentration until a non-rate-limiting excess of substrate is reached, after which additional
substrate causes no increase in rate. The region in
which linearity between reaction rate and substrate
concentration is achieved, and in which an analytical
determination of substrate concentration can be made
based on the rate of reaction, lies below $0.2 \; K_m$.
K_m is the Michaelis constant and is defined on p. 3.
The most important advantage of an enzymatic assay
is its selectivity. Frequently only one member of
a homologous series is active in the enzyme catalyzed reaction; other members are totally inactive
or react at much slower rates. Most enzymes are
also specific for one optical isomer of a substrate.
Thus in the enzymatic assay of amino acids, bacterial amino acid decarboxylase is specific for L-
amino acids.[1] Another advantage in the use of enzymes for substrate analysis lies in the great sensitivity obtained.

B. CARBOHYDRATES

1. Glucose

Glucose has undoubtedly received much attention

114

from analysts during the past 20 years. Of all the
methods proposed for the determination of glucose,
many are enzymatic, using either the enzyme hexokinase or glucose oxidase.

a. Determination with Hexokinase. Hexokinase
catalyzes the phosphorylation of glucose by the coenzyme ATP (adenosine triphosphate): an indicator
reaction is used to monitor the hexokinase reaction.
The enzyme glucose-6-phosphate dehydrogenase is used
in the presence of NADP (nicotinamide adenine dinucleotide phosphate):

Glucose + ATP $\xrightarrow{\text{Hexokinase}}$ glucose-6-phosphate + ADP

Glucose-6-phosphate + NADP $\xrightarrow{\text{Dehydrogenase}}$

$$\text{NADPH} + 6\text{-phosphoglucono-}\delta\text{-lactone} + H^+$$

The NADPH produced is monitored spectrophotometrically[2,3] or fluorometrically.

In the absence of contaminating enzymes this procedure is specific for glucose (and glucose-6-phos-
phate). Prior deproteinization with $Ba(OH)_2$ and
$ZnSO_4$ removes the latter and allows the specific determination of glucose. There are no interferences.

b. Determination with Glucose Oxidase. Glucose
oxidase catalyzes the reaction

$\beta\text{-D-Glucose} + O_2 + H_2O \xrightarrow[\text{Oxidase}]{\text{Glucose}} H_2O_2 + \text{D-gluconic acid}$

Enzymic methods using glucose oxidase (pp. 83-87)
have been found to be extremely sensitive[4] (as little
as 0.01 µg of glucose being determinable) and specific. A rather complete study of about 60 oxidizable
sugars and their derivatives showed that only 2-
deoxy-D-glucose is catalyzed at a rate comparable to
that of β-D-glucose. The anomer α-D-glucose is oxidized catalytically less than 1% as rapidly as the
β-anomer.[5]

Various methods have been proposed to monitor the
glucose concentration in the enzymatic glucose oxidase

procedure. The oxygen uptake has been monitored manometrically, or electrometrically with an oxygen electrode. Colorimetric, fluorometric and electrochemical indicator reactions have been used to measure the peroxide produced in the glucose oxidase catalyzed oxidation of glucose.

1. Measurements of Oxygen Uptake. One of the first methods described for the assay of glucose with glucose oxidase was a manometric one, in which the oxygen uptake was measured with a Warburg apparatus.[6]

Guilbault et al[7] described a sensitive electrochemical method for glucose (pp. 84-85). Calibration plots of ΔE/min vs glucose concentration were linear in the range of 2 to 42 μg per ml.

Alternatively the concentration of glucose can be determined by measuring the oxygen uptake using an oxygen specific electrode. This sensor consists of a gold cathode separated by an epoxy casting from a tubular silver anode. The inner sensor body is housed in a plastic casing and comes in contact with the process stream only through the Teflon membrane (Fig. 1). When oxygen diffuses through the membrane, it is electrochemically reduced at the cathode at an applied voltage of 0.8 volt. This reaction causes a current to flow between the anode and cathode which is proportional to the partial pressure of oxygen in the sample. Oxygen electrodes are available commercially (Beckman Instrument Co.) or can be made in the laboratory.[8]

Of the various techniques available for monitoring glucose in blood, many researchers feel the oxygen electrode method to be the most reliable. Kadish and Hall[9] and Makino and Koono[10] found a good correlation between glucose values determined in blood by a measurement of oxygen uptake with those found by standard chemical tests.

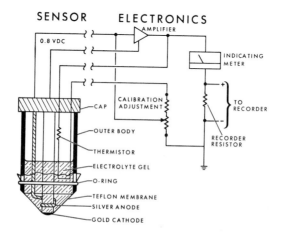

FIG. 1

Diagram of Oxygen Electrode

(Compliments of Beckman Instrument Co.)

Kadish, Litle and Sternberg[11] used the Beckman 777 polarographic oxygen sensor electrode with a circuit modified to record the rate of oxygen consumption as a measure of glucose levels in serum, plasma and urine. The maximum apparent rate of oxygen consumption relative to the rate obtained with a glucose standard provides a direct measure of the glucose level in the sample. The authors[11] were able to obtain results within 20 sec. after sample addition and within 3 minutes of sampling. As little as 0.1 ml of blood was analyzed with a deviation of less than 1.5% on replicate determinations and a bias of 1% with respect to data obtained on samples also run by the automated ferricyanide method.

Updike and Hicks[12] have prepared an enzyme electrode that responds to glucose in blood by placing a gelatinous membrane of immobilized enzyme over a polarographic oxygen electrode. The authors found an excellent agreement between their results and those obtained by an automated standard chemical method. This device will be discussed further in Chapter 7.

2. Use of Peroxidase. The enzyme peroxidase can be used in a coupled indicator reaction to indicate the amount of glucose present. The intensity of the colored or fluorescent dye produced is a measure of the concentration of glucose present. In the conventional Glucostat test, the dye o-dianisidine is used[13], although other dyes such as o-tolidine[14] and a β-diketone[15] have been used.

Guilbault, Brignac and Zimmer[4] proposed a fluorometric method for the sensitive determination of glucose in biological samples like blood and urine. The peroxide produced upon enzymic action is monitored with homovanillic acid, which is oxidized to a highly fluorescent product (pp. 86-87). The rate of production of fluorescence is proportional to the concentration of glucose in the concentration range 0.01-10µg. per ml (an increase in sensitivity of two orders over colorimetric methods). Phillips and Elevitch[16] used the HVA procedure for the assay of glucose in plasma; as little as 1 µl of sample was needed.

A complete survey of methods for the analysis of glucose can be found in reviews by Guilbault.[17,18] Interference in the glucose analysis will occur if the sample contains disaccharides and if the glucose oxidase preparation is contaminated with hydrolytic enzymes such as lactase, amylase or maltase which will liberate glucose from these disaccharides. Large amounts ($>$10 mg) of reducing agents such as ascorbic acid, uric acid, hydroquinone and bilirubin interfere by competing with peroxidase for the hydrogen peroxide formed.[19] For example, Hollister, Helmke and Wright[20] compared the blood glucose determined by an enzyme strip test (paper strips impregnated with glucose oxidase, peroxidase and leuco dye) with that found with the AutoAnalyzer in 542 cases and found that the strip method has limited usefulness and should only be used where other methods are not available.

3. Other Methods. Indicator reactions not using peroxidase have been described for the assay of glucose. Some of these methods (i.e., the electrochemical procedures of Pardue and Malmstadt[21-24] and Blaedel and Olson[25]) have been previously described in Chapter 3 (pp. 84,86).

Guilbault and coworkers have devised a fluorometric reaction to monitor the peroxide formed in the glucose-glucose oxidase enzymic reaction.[26] Phthalic anhydride(I) is used which reacts with peroxide to form a peroxy phthalate(II); this peroxy compound is a strong

oxidant and oxidizes indole(III) to the highly fluorescent indigo white(IV). The rate of production of
fluorescence is a measure of the glucose present.

II +

III
(Non-Fluorescent)

IV
(Highly Fluorescent)

    c. Determination with Other Enzymes. Bergmeyer
and Moellering[27] reported the enzymatic determination
of glucose with acyl phosphate and D-glucose-6-phospho-
transferase and found the method more specific than
the hexokinase procedure. Boehringer and Soehne[28]
used glucose transferase, acyl phosphate and glucose-
6-phosphate dehydrogenase to measure glucose in black-
berries and blood. Again the NADPH formed is measured and equated to glucose concentration.

2. Galactose

    Essentially the same methods described above with
glucose can be applied to the determination of galactose with the substitution of galactose oxidase for
glucose oxidase. The Galactostat reagent, for example,
is identical to the Glucostat, except for the enzyme
used.

    Guilbault, Brignac and Juneau[29] have developed
a fluorometric method for galactose using galactose
oxidase, in a procedure similar to that developed for
glucose with homovanillic acid. Because the lowest
detectable concentration of galactose was only 50 μg
using HVA, a systematic study was conducted for other
substrates which might offer better sensitivity using
an initial rate method. Of 25 substrates surveyed, 3
were found that work well with galactose in the 0.1-20μg

range: p-hydroxyphenylacetic acid, tyramine and tyrosine. These compounds are oxidized via a mechanism
similar to that for homovanillic acid (HVA), yielding
fluorescent products with fluorescence excitation and
emission wavelengths similar to oxidized HVA, but with
higher fluorescence coefficients (total fluorescence/
concentration in Molarity). All were stable in aqueous
solution, and cost considerably less than HVA. p-
Hydroxyphenylacetic acid(V), which was judged to be the
best substrate for oxidative enzymes, costs 0.44 cents
per gram compared to $30.00 per gram for HVA.

$$\text{Galactose} + O_2 \xrightarrow[\text{Oxidase}]{\text{Galactose}} H_2O_2$$

$$H_2O_2 + \;V\;(\text{Non-Fluorescent}) \longrightarrow \;VI\;(\text{Fluorescent})$$

Galactose oxidase is not as highly specific as
glucose oxidase. The C_1 position need not be free,
since galactosides are readily attacked. The β configuration is somewhat favored. However, this is not
an important structural requirement. The formula for
a number of common sugars are given below. These
formulae are written in straight chain rather than
cyclic for ease of comparison.

D-Glucose	D-Galactose	D-Talose	D-Lyxose	L-Altrose	D-Gulose
CHO	CHO	CHO		CHO	CHO
$-C^2$-OH	H-C^2-OH	HO-C-H	CHO	H-C-OH	H-C-OH
HO-C^3-H	HO-C^3-H	HO-CH	HO-C-H	HO-CH	H-C-OH
H-C^4-OH	HO-C^4-H	HO-C-H	HO-CH	HO-C-H	HO-CH
H-C^5-OH	H-C^5-OH	H-C-OH	H-C-OH	HO-C-H	H-C-OH
H_2-C^6-OH	H_2-C^6-OH	H_2-C-OH	H_2-C-OH	H_2-C-OH	H_2-C-OH

The galactose configuration at position 4 is essential; glucose and its derivatives are completely
inert. The configuration at position 2 is not so

critical since D-talose, 2-deoxy-D-galactose, and D-galactosamine are good substrates, comparable to D-galactose.[30]

It was found in this study that if the number of carbon atoms is reduced by one, keeping the same configuration as D-galactose (for example, D-lyxose), the rate of oxidation is reduced drastically. Also L-altrose, which differs from D-galactose only in the C_5 position, is oxidized at a rate comparable to D-galactose. Thus the C_5 position is not so critical. The C_3 position is critical because D-gulose, which differs from D-galactose only at this position, is not oxidized at all. From these structural considerations, the enzyme can be very useful as a selective reagent in sugar analyses. The following sugars are catalytically oxidized by galactose oxidase and can be determined in the 0.1-20 μg region: stachyose, 2-deoxy-D-galactose, methyl-β-D-galacto-pyranoside, D-raffinose, D-galactosamine, N-acetyl-D-galactosamine and α-D-melibiose.

3. Fructose

Fructose can be analyzed using the enzymes hexokinase, phosphoglucose isomerase and glucose-6-phosphate dehydrogenase[31]:

$$\text{Fructose} + \text{ATP} \xrightarrow{\text{Hexokinase}} \text{Fructose-6-phosphate} + \text{ADP}$$

$$\text{Fructose-6-phosphate} \xrightarrow[\text{isomerase}]{\text{Phosphoglucose}} \text{glucose-6-phosphate}$$

$$\text{Glucose-6-phosphate} + \text{NADP} \xrightarrow{\text{Dehydrogenase}}$$
$$\text{NADPH} + \text{H}^+ + \text{6-phospho-δ-lactone}$$

The NADPH produced is measured colorimetrically and is a measure of the fructose present. All three reactions proceed stoichiometrically. The reaction sequence is

highly selective for fructose. Mannose can be phosphorylated by hexokinase but the mannose-6-phosphate
produced does not react further.

Guilbault and co-workers[32] have proposed a fluorometric assay of fructose, using the resazurin-
resorufin reaction. The NADPH produced in the glu-
cose-6-phosphate reaction above effects the reduction
of the non-fluorescent resazurin to the highly fluorescent resorufin in the presence of PMS (phenazine
methyl sulfate). The rate of production of resorufin
is proportional to the amount of fructose in concentrations as low as $1.7 \times 10^{-6} \underline{M}$ (0.3 µg/ml).

4. Sucrose

Sucrose is hydrolyzed by invertase to D-glucose
and fructose

$$\text{Sucrose} \xrightarrow{\text{Invertase}} \text{D-glucose} + \text{fructose}$$

The glucose produced can be measured in an indicator reaction with glucose oxidase colorimetrically
or fluorometrically, or with hexokinase and glucose-
6-phosphate dehydrogenase:

$$\text{Glucose} + \text{ATP} \xrightarrow{\text{Hexokinase}} \text{glucose-6-phosphate}$$

$$\text{NADPH} + \text{glucose-6-phosphate} \xrightarrow{\text{Dehydrogenase}}$$

$$\text{NADPH} + \text{6-phosphogluconate}$$

The NADPH produced is measured colorimetrically or
fluorometrically.

Guilbault et al[29] applied the fluorometric p-
hydroxyphenylacetic acid procedure described above
to the determination of sucrose. Sucrose is first
hydrolyzed to glucose by incubation with invertase
at pH 6, then after 10 minutes the glucose
produced is analyzed fluorometrically with glucose oxidase, peroxidase and p-hydroxyphenylacetic acid. The rate of production of fluorescence

is proportional to the concentration of sucrose in the range 2-100 µg/ml. Glucose is an interference.

Guilbault et al[29] developed a procedure for analysis of a 3 component mixture of β-D-glucose, D-galactose and sucrose using 3 enzyme systems: glucose oxidase, galactose oxidase and invertase. Analysis was possible because of the specificity built into these systems. It was found that concentrations of galactose up to 100 times that of glucose did not interfere in the determination of the latter, and vice versa. Three component mixtures of glucose, galactose and sucrose were analyzed for all three components with an accuracy and precision of about 1.5%. One aliquot (A) was analyzed for glucose using glucose oxidase, another (B) for galactose using galactose oxidase in procedures as described above in sections 1 and 2.   A third aliquot (C) was analyzed for sucrose by addition of invertase to liberated glucose, followed by a determination of total glucose with glucose oxidase. The amount of sucrose present was calculated by subtracting the glucose found in A from that found in C.

5.   Glycogen

Glycogen is a polymer consisting of several glucose units. Glycogen can be determined by hydrolysis to glucose, which can then be determined by any of the standard glucose methods. In the procedure of Pfleiderer and Grein[33] the enzymes hexokinase, pyruvate kinase and lactic dehydrogenase are used:

$$\text{Glycogen} \xrightarrow{H^+} \text{Glucose}$$

$$\text{D-glucose} + \text{ATP} \xrightarrow[\text{Mg}^{++}]{\text{Hexokinase}} \text{D-glucose-6-phosphate} + \text{ADP}$$

$$\text{ADP} + \text{Phosphoenolpyruvate} \xrightarrow[\text{Mg}^{2+}]{\text{Pyruvate Kinase}}$$
$$\text{ATP} + \text{Pyruvate}$$

$$\text{Pyruvate} + \text{NADH} + \text{H}^+ \xrightarrow{\frac{\text{Lactate}}{\text{Dehydrogenase}}} \text{lactate} + \text{NAD}$$

The decrease in NADH is a measure of the glycogen content of the sample.

Rerup and Lundquist[34] determined glycogen using the glucose oxidase-peroxidase-o-tolidine reaction to monitor the glucose produced. Passonneau et al[35] described an enzymatic method for glycogen based on the measurement of NADPH produced in the following enzyme sequence:

$$\text{Glycogen} + \text{phosphate} \xrightarrow{\text{phosphorylase}} \text{glucose-1-phosphate}$$

$$\text{Glucose-1-phosphate} \xrightarrow{\text{phosphoglucomutase}} \text{glucose-6-phosphate}$$

$$\text{Glucose-6-phosphate} + \text{NADP} \xrightarrow{\text{Dehydrogenase}}$$

$$\text{NADPH} + \text{6-phospho-gluconolactone} + \text{H}^+$$

Glycogen in as little as 30 μg of brain or 0.3 μg of liver can be specifically determined.

6.   Other Sugars

Raffinose (the trisaccharide of glucose, fructose and galactose) can be analyzed directly with galactose oxidase by a fluorometric procedure developed by Guilbault et al[29] (see pp. 121,122 for details). Or raffinose can be hydrolyzed by invertase to fructose and melibiose. The melibiose produced is hydrolyzed to glucose and galactose by melibiase. Thus the raffinose can be determined either by an enzymic assay of glucose or

$$\text{Raffinose} \xrightarrow{\text{invertase}} \text{fructose} + \text{melibiose}$$

$$\text{Melibiose} \xrightarrow{\text{melibiase}} \text{glucose} + \text{galactose}$$

galactose, by assay of the reducing sugars formed with Fehling's solution, or by means of a change in the optical rotation. Of all methods, the fluorometric one of Guilbault is the most sensitive (0.1-50 μg determinable) and the fastest (2-3 minutes for complete analysis).

Lactose (a disaccharide of glucose and galactose) is hydrolyzed by β-galactosidase to galactose and glucose:

$$\text{Lactose} + H_2O \xrightarrow{\text{β-galactosidase}} \text{galactose} + \text{glucose}$$

Guilbault and coworkers coupled this reaction to a glucose oxidase-peroxidase-p-hydroxyphenylacetic acid indicator reaction, and were able to fluorometrically[32] determine 0.1-50 μg per ml of lactose in about 15 minutes.

Likewise lactose can be assayed by noting the increase in absorbance of NADPH formed using the hexokinase, glucose-6-phosphate dehydrogenase mixed enzyme system to assay the glucose formed from lactose.[36]

$$\text{Glucose} + \text{ATP} \xrightarrow{\text{Hexokinase}} \text{glucose-6-phosphate} + \text{ADP}$$

$$\text{Glucose-6-phosphate} + \text{NADP} \xrightarrow[\text{Dehydrogenase}]{\text{G6P}}$$

$$\text{6-phosphogluconate} + \text{NADPH}$$

Gluconate can be assayed specifically using the enzymes gluconokinase and 6-phosphogluconic dehydrogenase:

$$\text{Gluconate} + \text{ATP} \xrightarrow{\text{Gluconokinase}} \text{ADP} +$$
$$\text{6-phosphogluconate}$$

$$\text{6-Phosphogluconate} + \text{NADP} \xrightarrow{\text{Dehydrogenase}}$$

$$\text{NADPH} + CO_2 +$$
$$\text{ribose-5-phosphate}$$

The NADPH produced is measured colorimetrically and is a measure of the gluconate present.[37,38]

C. AMINES

1. Use of Diamine Oxidase

Diamine oxidase, an enzyme found in hog kidney, cat kidney, guinea pig and rabbit liver, human placenta, cattle and pig plasma, pea seedlings and Mycobacterium smegmates, catalyzes the deamination of amines to ammonia and hydrogen peroxide. A good review on the properties, kinetics and specificity of this reaction has been written by Zeller.[39]

The enzyme from hog kidney (available commercially from various companies) acts upon various alkyl and aryl amines, such as benzylamine, tyramine, mescaline, histamine and cadaverine. High concentrations of substrate and large quantities of enzyme are required for deamination of monoamines, the K_m values of diamines are about 1/10 those of the corresponding monoamines.[40]

For alkylamines a chain length of 4-5 carbons separating the amino groups is required (n=4 or 5) for easy degradation.

$$R-\underset{\underset{R'}{|}}{N}-C_n-NH_2$$

One amine group generally must be unsubstituted but the second amine function may be mono- or di-substituted (R or R' can be CH_3 or H).[39, 40]

Aromatic systems containing nitrogen can represent the second amine group; the aminoethyl deviatives of imidazole, pyrazole and 1,2,3-triazole, for example, are good substrates.[41] The 2- and 4-(2-aminoethyl) pyridines are not substrates, however.

Analytical methods for the analysis of amines using diamine oxidase have involved measurement of:(1) unreacted amine, either colorimetrically or fluorometrically[42,43]; (2) the oxygen consumption manometrically; (3) the aldehyde production, either directly(λ_{max} of

benzaldehyde formed from benzylamine is 280 mμ) or
by formation of a Δ'-pyroline derivative   (λ_{max} $\sim$
430 mμ)[44,45]; (4) the ammonia liberated manometrically[46,47]; or (5) the hydrogen peroxide formed with a
peroxidase dye reaction similar to those already discussed above. (See determination of peroxidase pp. 87-89).

Guilbault and co-workers have used a Beckman 39137
cation selective electrode for the assay of amines. The
electrode responds to the ammonium ion produced linearly
over the concentration range 10^{-1} to $10^{-5}\underline{M}$. The rate of
change in the potential with time due to ammonium ion
production, is proportional to the concentration of amine
present at concentrations of 1-100 μg/ml.[48]

2.   Use of Monoamine Oxidase

The function of monoamine oxidase in the animal body
is the decomposition of biologically important amines.
Many naturally occuring amines and pharmacologically
important amines are substrates in vivo. Similarly,
monoamine oxidase can be used for the in vitro determination of amines.[49]

Primary and secondary amines are readily attacked by
monoamine oxidase; only methyl substituted amines are
substrates. The rate of oxidation of some tertiary
amines is high with some enzymes (e.g. cat), low with
others (e.g. rabbit). The enzyme does not react with
methylamine, reacts slowly with ethylamine, and maxi-
mally with amyl- or hexyl-amine (n = 4-5 in the homologous series CH_3 $(CH_2)_n NH_2$). In contrast to diamine
oxidase, the diamines putrescine and cadaverine are not
attacked.

Most natural substrates of monoamine oxidase are
cyclic, such as phenylethylamine, tryptamine, histamine,
dopamine, norepinephrine, epinephrine and tyramine.[49]

The same methods described for the assay of substrates
of diamine oxidase can be used for the assay of

the substrates of monoamine oxidase.

Guilbault and Brignac[50] have developed a fluorometric method for benzylamine, furfurylamine, tyramine,
histamine and other amines. The peroxide produced in
the monoamine oxidase reaction is detected with peroxidase and p-hydroxy phenylacetic acid(V).

$$\text{Benzylamine} \xrightarrow{\frac{\text{Monoamine}}{\text{Oxidase}}} \text{Benzaldehyde} + H_2O_2 + NH_4^+$$

The rate of formation of fluorescence due to production
of oxidized p-hydroxy phenylacetic acid(VI) is proportional to the concentration of benzylamine.

Bachrach and Reches[51] assayed spermine and sper-
midine using an amine oxidase catalyzed oxidation, and
McEqen and Sober[52] studied the interaction of primary,
aliphatic amines with highly purified rabbit serum monoamine oxidase. An enzymatic method for the determination
of the racemization rate of hyoscyamine and scopolamine
was described by Werner and Seiler.[53]

3.   Urea Analysis with Urease

Urease can be used for the specific assay of urea,
the reaction being followed either manometrically by
noting the CO_2 evolution, or by measurement of the NH_3
liberated either electrometrically or colorimetrically
(acid-base indicator).

One of the most common methods of assay involves the
use of the Nessler reaction to detect the ammonia produced. Nessler's reagent (K_2HgI_4) reacts with ammonia
to give a colored product ($NH_2Hg_2I_3$) (λ_{max} = 436 mμ).[54]
This reaction, though convenient, suffers from the serious
disadvantage of the instability of the Nessler's reagent
(must be prepared fresh hourly). Ammonia can be determined by an acid base titration after distillation[55]
or after diffusion.[56] Such techniques are undesirable,
particularly in a micro determination.

Other colorimetric methods have been proposed for the analysis of the ammonia liberated. Naftalm, Whitaker and Stephens[57] used the absorbance change at 660 mμ after addition of hypobromite to measure the ammonia produced from urea in blood. Wilson[58] described an automatic method for the determination of urea using urease, hypochlorite and alkaline phenol that offers advantages of speed and precision over the conventional Nessler's reagent. Cirje and Sandru assayed urea in blood and urine using this improved method.[59] An enzyme chromatographic method (Urastat strip test) was used for the determination of blood urea nitrogen levels in ox, horse, pig and sheep by Parmense[60], and a comparison of the Urastat method with the xanthydrol hypobromite micromethod for urea in blood by Manzini[61] showed the former had greater simplicity, speed and reliability.

A coupled optical enzyme assay for urease was developed by Kaltwasser and Schlegel.[62] The NADH dependent glutamate dehydrogenase was used; the rate of ammonia production from urease was calculated from the rate of NADH oxidation (disappearance of absorbance at 340 mμ):

$$\text{Urea} \xrightarrow{\text{Urease}} NH_4^+$$

$$NH_4^+ + \alpha\text{-ketoglutarate} + NADH \xrightarrow{\substack{\text{Glutamate} \\ \text{Dehydrogenase}}}$$

$$\text{Glutamic acid} + NAD$$

This same method was used by Roch-Ramel[63] except that the NAD formed from NADH was measured fluorometrically. From 2×10^{-11} to 10^{-10} mole of urea was determined.

Several electrochemical procedures have been proposed for the assay of urea using urease. Nielsen[64] deter-

mined urea in blood and urine using a pH meter; Malmstadt
and Piepmeier[65] used a pH stat with digital readout.
Purdy, Christian and Knoblock[66] followed the ammonia
produced in the enzymic reaction coulometrically.

Katz[67] and Katz and Rechnitz[68] have described a
potentiometric method for urease. A Beckman cationic
sensitive glass electrode that responds to $[NH_4^+]$ is
used to follow the course of the reaction. Guilbault
et al[48] used an ammonium ion selective electrode for
the automatic assay of urea and urease. The rate of
change in the potential of the electrode $\Delta E/min$, is proportional to the concentration of urea over the range
0.1-50 $\mu g/ml$.

D. AMINO ACIDS

1. Determination of D-Amino Acids

D-Amino acid oxidase, found in the kidney and liver
of all mammals especially the sheep and pig, catalyzes
the deamination of D-amino acids:

$$R-\underset{\underset{NH_2}{|}}{CH}-COOH + O_2 \longrightarrow R-\underset{\underset{\downarrow}{\overset{\|}{NH}}}{C}-COOH + H_2O_2$$

$$R-\underset{\overset{\|}{O}}{C}-COOH + NH_3$$

D-amino acids, which are catalytically oxidized by this
enzyme, can thus be specifically determined in the
presence of L-amino acids and unreactive amino acids.
Some of the substrates of D-amino acid oxidase from
sheep kidney[69] are listed in Table 2. Several methods
have been proposed for analysis of amino acids using
amino acid oxidase: manometric determination of oxygen
uptake using a Warburg apparatus; reaction of the α-keto
acid formed with o-phenylenediamine (Wieland's reagent)
using UV analysis of the yellow product[70], analysis

TABLE 2

Relative Rates of Oxidation of D-amino Acids by D-amino
Acid Oxidase from Sheep Kidney.   Tyrosine = 100

D-Amino Acid	Relative Rate
Tyrosine	100
Proline	78
Methionine	42
Alanine	34
Serine	22
Tryptophan	19.5
Valine	18.4
Phenylalanine	13.7
Isoleucine	11.6
Leucine	7.4
Histidine	3.3
Glutamic Acid	0

of the peroxide formed with any of the conventional
chromogenic indicators, or electrochemical determination
of the ammonia produced with an ammonium ion selective
electrode. [48]

Guilbault and Hieserman[71] have proposed a fluorometric assay procedure for the assay of the D-amino
acids.   The peroxide formed oxidized the non-fluorescent
homovanillic acid to the highly fluorescent 2,2'-dihy-
droxy 3,3'-dimethoxy biphenyl 5,5'-diacetic acid in
the presence of peroxidase.   The initial rate of formation of this fluorescent compound is measured and
related to the activity of the D-amino acids: alanine,
methionine,phenylalanine, proline, tryptophan, tyrosine
and leucine in the 1-100 μg/ml region (Figure 2 and
Table 3).   The enzyme used, from hog kidney has an
order of specificity similar to that from a sheep kidney.

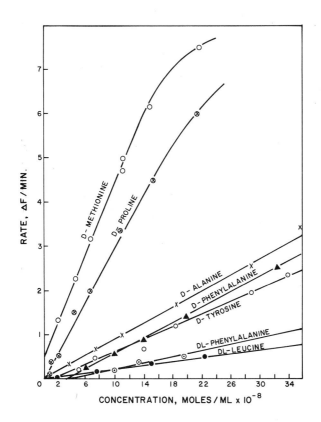

FIG. 2

Variation of ΔF/min with concentration of various
D-amino acids using D-AAO
(ref. 71)

TABLE 3

D- and L-Amino Acids Determined with
Amino Acid Oxidase (AAO)

Name	Av. Rel. Error %	Useful Range (μg/ml.)
Determined with D-AAO		
D-alanine	± 2.7	1 - 150
D-methionine	± 2.6	1 - 36
D-phenylalanine	± 1.3	1 - 60
D-proline	± 2.3	1 - 40
D-tryptophan	± 1.7	10 - 200
D-tyrosine	± 3.1	1 - 40
D-leucine	± 1.8	5 - 250
Av. Rel. Error	± 2.2	
Determined with L-AAO		
L-arginine	± 1.8	0.4 - 40
L-leucine	± 1.8	0.1 - 50
L-methionine	± 2.4	0.03- 8.0
L-phenylalanine	± 2.4	0.01- 5.0
L-proline	± 3.0	10 - 200
L-tryptophan	± 1.6	0.03- 12
L-tyrosine	± 3.3	0.05- 50
DL-phenylalanine	± 0.49	0.3 - 25
Av. Rel. Error	± 2.0	

D-valine, D-histidine, D-α-aminobutyric, D-α-amino-
valeric acid, D-aspartate and D-threonine also react
and are determinable. Mixtures of amino acids can
be easily separated by ion exchange prior to analysis.

2. Determination of L-Amino Acids with L-Amino Acid
    Oxidase.

L-Amino acid oxidase, from snake venom, effects
the deamination of L-amino acids in much the same
way as D-amino acid oxidase.

$$\text{L-amino acid} + O_2 \xrightarrow{\text{L-AAO}} \alpha\text{-keto acid} + NH_3 + H_2O_2$$

The same analytical methods described for D-amino
acid oxidase above are applicable to the L-amino acid
oxidase system: manometric measurement of the oxygen
uptake, electrochemical measurement of the NH_3 produced with a cation selective electrode[48] or colorimetric or fluorometric monitoring of the peroxide
formed.

Guilbault and Hieserman[71] found that 0.01-5.0 µg
per ml of L-arginine, -leucine, -methionine, -phenyl-
alanine, -proline, -tryptophan, and -tyrosine could
be measured using the fluorometric peroxidase-homo-
vanillic acid monitoring system (Fig. 3), with an
average relative error of about 2% (Table 3). The
order of reactivity of L-amino acids with snake venom
was L-phenylalanine > -methionine>-tryptophan >-tyrosine>
-leucine >-arginine >-proline. L-α-Aminovaleric acid
-α-aminocaproic acids, and p-substituted L-phenyl-
alanines are determinable. No other L-amino acids
were found to be reactive under conditions of this
assay. Mixtures of L-amino acids must be separated
by ion exchange prior to analysis.

3. Determination of L-Amino Acids with L-Amino Acid
    Decarboxylases

L-amino acid decarboxylases, grown from certain
bacteria under specific conditions[72,73], catalyze
reactions of the type:

$$\text{R-CH-COOH} \longrightarrow \text{R-CH}_2\text{-NH}_2 + CO_2$$
$$\quad\quad |$$
$$\quad NH_2$$

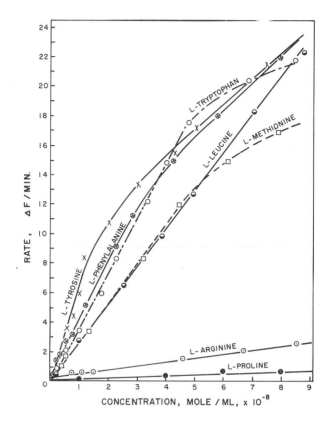

FIG. 3

Variation of initial rate of reaction, ΔF/min, as
function of concentration of various L-amino
acids using L-AAO

(ref. 71)

The carbon dioxide liberated is measured manometrically using a Warburg apparatus.[72,73]

Specific decarboxylases are available for the following acids:

L-tyrosine $\xrightarrow{\text{E}}$ tyramine + CO_2

L-histidine $\xrightarrow{\text{E}}$ histamine + CO_2

L-ornithine $\xrightarrow{\text{E}}$ putrescine + CO_2

L-lysine $\xrightarrow{\text{E}}$ cadaverine + CO_2

L-arginine $\xrightarrow{\text{E}}$ agmatine + CO_2

L-glutamic acid $\xrightarrow{\text{E}}$ γ-aminobutyric acid + CO_2

L-aspartic acid $\xrightarrow{\text{E}}$ α-alanine + CO_2

The pH optimum for most of these enzymes is between 4.5 and 6. Each enzyme preparation is specific for its respective L-amino acid substrate. Lysine decarboxylase, however, may contain traces of arginine decarboxylase which can be removed by keeping the acetone dried lysine decarboxylase preparation for 2-3 days at 0°C. Similarly, traces of glutamic decarboxylase impurities in histidine decarboxylase can be removed by overnight incubation of the acetone dried powder in pH 8.5 borate buffer at 37°C.

Tyrosine decarboxylase attacks phenylalanine[74] at a rate about 1/15 that of tyrosine; lysine decarboxylase will also decarboxylate hydroxylysine, though at a slower rate.

Guilbault et al have proposed a fluorometric assay procedure for L-amino acids using L-amino acid decarboxylases. The amine produced in the primary enzyme reaction is determined with an amine oxidase and a peroxidase-p-hydroxyphenylacetic acid(V) mixed indicator reaction.

The rate of production of the highly fluorescent compound (VI), ΔF/min, is proportional to the concentration of amino acid present. [50]

4.   Determination of L-Glutamic Acid

   a. Use of Glutamate Dehydrogenase. Glutamate dehydrogenase catalyzes the reaction:

$$\text{L-glutamate} + \text{NAD} + H_2O \longrightarrow \alpha\text{-oxoglutarate} + \text{NADH} + NH_4^+$$

The rate of production of NADH is measured colorimetrically and is proportional to the concentration of glutamic acid.

   Since the equilibrium of this reaction lies to the left, several steps must be taken to help shift the equilibrium: hydrazine is added, which traps the α-oxoglutarate by hydrazone formation; an alkaline medium (pH 9-10) is used; and excess NAD is added.

   Guilbault et al[75,76] have proposed a more sensitive fluorometric method for monitoring this reaction. The dye resazurin is used which is non-fluorescent, but is converted to the highly fluorescent resorufin by NADH in the presence of an electron transfer reagent phenazine methosulfate (PMS). The rate of production of resorufin with time, ΔF/min, is proportional to the L-glutamate concentration.

   Guilbault et al[76] found that if hydrazine is not added to complex the α-oxoglutarate, the reaction with glutamate dehydrogenase is completely specific for L-glutamate. With hydrazine present, the enzyme catalyzes oxidations with rates in the order L-glutamic $\rangle$ D-isocitric $\rangle$ L-malic $\rangle$ glycolic $\rangle$ oxalic $\rangle$ L-lactic $\rangle$ DL-citric acid.

   An alternative to the use of a hydrazine to shift the equilibrium of the glutamate-glutamate dehydrogenase reaction, is to use the 3-acetylpyridine analog of NAD (AP-NAD)[77], which has a more favorable redox potential.

$$\text{L-glutamate} + \text{AP-NAD} + \text{H}_2\text{O} \rightleftharpoons \alpha\text{-oxoglutarate} + \text{AP-NADH} + \text{NH}_4^+$$

The formation of AP-NADH is measured at 366 mμ and is equated to the concentration of L-glutamate. This reaction is almost completely specific for L-glutamate; very high concentrations of serine and norvaline interfere.

b. <u>Use of Glutamate-Pyruvate Transaminase</u>. Glutamic acid can be determined with the enzyme glutamate-pyruvate transaminase (GPT), which transfers the α-amino group of glutamic acid to pyruvate:

$$\text{L-glutamate} + \text{Pyruvate} \xrightarrow{\text{GPT}} \alpha\text{-oxoglutarate} + \text{L-alanine}$$

$$\alpha\text{-oxoglutarate} + \text{NH}_4^+ + \text{NADH} \xrightarrow{\text{Glutamate}}_{\text{Dehydrogenase}}$$

$$\text{L-glutamate} + \text{NAD}$$

The decrease in NADH is measured spectrophotometrically at 340 or 366 mμ.

This reaction is completely specific for L-glutamic acid; there are no interferences. The transaminase enzyme is very expensive, however.

5. Determination of L-Alanine

Glutamate-pyruvate transaminase (GPT) can also be used for the determination of L-alanine:

$$\text{L-Alanine} + \alpha\text{-oxoglutarate} \xrightarrow{\text{GPT}} \text{L-glutamate} + \text{pyruvate}$$

The pyruvate produced can be followed using lactate dehydrogenase (LDH) and NADH:

$$\text{Pyruvate} + \text{NADH} + \text{H}^+ \xrightarrow{\text{LDH}} \text{L-lactate} + \text{NAD}$$

The disappearance of NADH is measured spectrophotometrically. With an excess of NADH, GPT and LDH the rate of decrease in absorbance is proportional to the concentration of L-alanine. This reaction is completely specific for L-alanine in the 50-500 µg range.

6.   Determination of L-Aspartic Acid

Glutamate-oxaloacetic transaminase (GOT) catalyzes the conversion of aspartate and α-oxoglutarate to oxaloacetate and glutamate:

$$\alpha\text{-Oxoglutarate} + \text{L-aspartate} \xrightleftharpoons{\text{GOT}} \begin{array}{l} \text{L-glutamate} \\ + \text{ oxaloacetate} \end{array}$$

The oxaloacetate produced is measured with malate dehydrogenase (MDH) and NADH:

$$\text{Oxaloacetate} + \text{NADH} + \text{H}^+ \xrightleftharpoons{\text{MDH}} \text{L-malate} + \text{NAD}$$

Again the decrease in NADH can be followed spectrophotometrically.[78,79] Using an excess of GOT, MDH and NADH, the rate of decrease in absorbance at 340 mµ is proportional to the L-aspartate concentration.

This method is highly selective; only cysteric acid interferes.

7.   Determination of Creatine

Creatine, methylguanidinoacetic acid, can be specifically determined by phosphorylation with ATP and creatine phosphokinase (CPK) to creatine phosphate:

$$\text{Creatine} + \text{ATP} \xrightleftharpoons{} \text{Creatine phosphate} + \text{ADP}$$

A dual indicator reaction using phosphoenolpyruvate (PEP) and pyruvate kinase (PK), and lactate dehydrogenase (LDH) is used to monitor the creatine phosphate formed:

$$\text{ADP} + \text{PEP} \xrightleftharpoons{} \text{ATP} + \text{Pyruvate}$$

$$\text{Pyruvate} + \text{NADH} + \text{H}^+ \xrightleftharpoons{} \text{Lactate} + \text{NAD}$$

The decrease in NADH, measured spectrophotometrically at

340 or 366 mµ, is proportional to the amount of creatine. The reaction is completely specific for creatine. There are no interferences.[80]

E. ORGANIC ACIDS

1. Acetic Acid

Acetate can be determined with a pigeon liver enzyme preparation which catalyzes the reaction[81]:

Acetate + ATP ———> acetyl-AMP + pyrophosphate

The reaction is monitored using coenzyme A (CoA) in the reaction sequence:

Acetyl-AMP + CoA ———> acetyl-CoA + AMP

Acetyl-CoA + Sulfanilamide ———>acetylsulfanilamide
                                      + CoA

The unreacted sulfanilamide is determined by the method of Bratton and Marshall[82], using sulfamic acid and N-naphthyl-ethylenediamine. The absorbance at 540 mµ is measured.

This procedure is highly selective for acetate. Propionate reacts at a rate about 1/150 that of acetate, butyrate about 1/80 as fast. There is a disadvantage to this procedure: the reaction sequence described is not stoichiometric; a yield of about 85% is obtained, but the total % conversion depends on the concentration of enzyme.

Guilbault, McQueen and Sadar[76] have proposed the use of malic dehydrogenase, in conjunction with a coupled fluorescence indicator reaction for the assay of 10-200 µg/ml concentrations of acetic acid. The rate of production of resorufin with time, $\Delta F/min$, is proportional to the acetic acid present.

Acetic acid + 2 NAD + 6 OH⁻ $\xrightarrow{\text{Malate Dehydrogenase}}$ 2 NADH + 2 CO_2 + 4 H_2O

NADH + Resazurin $\xrightarrow{\text{Phenazine Methyl Sulfate}}$ Resorufin + NAD

(Non-Fluorescent)   (Fluorescent)

Malic dehydrogenase (pig heart, Type I) was found to be a very non-specific enzyme system, catalyzing the oxidation of many α-hydroxy and non-hydroxy acids. The relative rates of oxidation of a number of acids are indicated in Table 4. In addition to acetic acid, adipic (110-2,500 µg/ml), benzilic (5-500 µg/ml), chloroacetic (100-2,500 µg/ml), DL-citric (150-600 µg/ml), formic (10-500 µg/ml), glutaric (100-2,500 µg/ml), glycolic (20-400 µg/ml), D-isocitric (20-400 µg/ml), L-malic (0.1-2.0 µg/ml), D-malic (50-800 µg/ml), malonic (20-550 µg/ml), oxalic (30-600 µg/ml), phthalic (50-1,000 µg/ml), DL-succinic (50-2,000 µg/ml) and D-tartaric (200-3,000 µg/ml) acids are determinable (Table 5).

2.   Butyric Acid and its Hydroxy Derivatives

Butyric acid can be determined using malate dehydrogenase in the 10-250 µg/ml range. Since most organic acids interfere, a prior separation of butyric acid must be effected.

Lactate dehydrogenase, Type IV from Bakers Yeast, is an enzyme which does not require the use of NAD or NADH. Electron acceptors such as ferricyanide or methylene blue can be used, and are directly reduced. Guilbault et al[76] proposed a fluorometric method for assay of this enzyme using resazurin as the electron acceptor. Resorufin is formed, and the rate of change in fluorescence per minute is proportional to the amount of enzyme or substrate present:

Lactate + Resazurin $\xrightarrow{\text{Type IV LDH}}$ Resorufin + Pyruvate

(Non-Fluorescent)   (Fluorescent)

TABLE 4

Relative Rates[a] of Lactic, Glutamic, and Malic
Dehydrogenase Systems

Acid = 100 µg/ml;   Enzyme = 0.03 mg/ml

Acid	LDH-II System[b]	GDH System[c]	MDH System[d]
Acetic	0	0	0.10
Adipic	0	0	0
Benzilic	0	0	0.12
Butyric	0	0	0.11
D-α-Hydroxy-Butyric	0	0	0.02
D-β-Hydroxy-Butyric	0	0	0.035
Chloroacetic	0	0	0.05
DL-Citric	0.03	0.04	0
Formic	0	0	0.15
L-Glutamic	0.16	0.35	0
Glutaric	0	0	0
Glycolic	0.53	0.12	0.18
D-Isocitric	0.16	0.18	0.23
L-Lactic	0.25	0.070	0
L-Leucine	0	0	0
L-Malic	0.01	0.13	0.25
D-Malic	0	0	0.052
Malonic	0	0	0.12
Oxalic	0.20	0.10	0.06
Phthalic	0	0	0.033
Salicylic	0	0	0
DL-Succinic	0	0	0.04
D-Tartaric	0	0	0
Tryptophan, Tyrosine	0	0	0

[a] Expressed as $\Delta F/min$.

[b] Glycine-Hydrazine buffer; with tris buffer only L-lactic reacts.

[c] Glycine-Hydrazine buffer; with tris buffer only L-glutamic acid reacts.

[d] Glycine-Hydrazine buffer.

TABLE 5

Lists of Acids Determinable Using Enzymes

Acid	Enzyme	Range, µg/ml	Interferences
Acetic	Acetate Kinase*	2 - 16	Propionic (>1 mg), Butyric (>1 mg)
	MDH	10 - 200	Adipic (>0.1 mg), butyric, D-α-OH and β-OH butyric (>0.1 mg), chloroacetic (>0.1 mg), DL-citric (>0.15 mg), formic, glutaric, glycolic, D-isocitric, L-malic, oxalic, phthalic, succinic, D-tartaric (>0.2 mg)
Adipic	MDH	110 - 2,500	Same as Acetic
Benzilic	MDH[a]	5 - 500	Same as Acetic
Butyric	MDH	20 - 250	Same as Acetic
D-α-OH-Butyric	MDH	100 - 1,500	Same as Acetic
	LDH-IV*	2 - 50	L-Lactic Acid
D-β-OH-Butyric	β-OH-BuDH	1 - 75	DL-Citric (>10 µg)
Chloroacetic	MDH	100 - 2,500	Same as Acetic
DL-Citric	MDH	150 - 600	Same as Acetic
	LDH-II	50 - 500	L-Glutamic, Glycolic, D-Isocitric, L-Lactic, L-Malic (>0.3 mg), Oxalic
	GDH	25 - 300	L-Glutamic, Glycolic, D-Isocitric, L-Lactic (>0.1 mg), L-Malic, Oxalic (>0.1 mg)
	β-OH-BuDH*	10 - 110	β-OH-Butyric
	Citrase*	1 - 10	Glutamic, α-ketoglutarate

TABLE 5 (Continued)

Substrate	Enzyme	Range (0.02–1.0)	Interferences
D-Isocitric	ICDH*	0.02 – 1.0	None
	GDH, LDH-II	10 – 250	Same as Citric
	MDH	20 – 400	Same as Acetic
Formic	MDH	10 – 500	Same as Acetic
	THFF	1 – 20	None
Fumarate	Fumarase	10 – 100	Pyruvate and those listed for MDH
L-Glutamic	GDH	1 – 100	None[b]
Glutaric	MDH	100 – 2,500	Same as Acetic
Glycolic	GDH	20 – 300	Same as Citric
	LDH-II*	2 – 75	Same as Citric
	MDH	20 – 400	Same as Acetic
Hyaluronic	Hyaluronidase*	50 – 1,000	None
L-Lactic	LDH-II	4 – 175	None[b]
	LDH-IV*	0.2 – 10	α-OH-Butyric
L-Malic	GDH	10 – 600	Same as Citric
	MDH*	0.1 – 2.0	Same as Acetic
D-Malic	MDH	50 – 800	Same as Acetic
Malonic	MDH	20 – 550	Same as Acetic
Oxalic	MDH	30 – 600	Same as Acetic
	LDH-II*	5 – 100	Same as Citric
Phthalic	MDH	50 – 1,000	Same as Acetic

F

TABLE 5 (Continued)

Acid	Enzyme	Range, μg/ml	Interferences
DL-Succinic	MDH	50 - 2,000	Same as Acetic
	SDH*	5 - 100	None
D-Tartaric	MDH	200 - 3,000	Same as Acetic

Legend:

GDH = glutamate dehydrogenase;   LDH = lactate dehydrogenase,   II = Type II,
IV = Type IV;   MDH = malic dehydrogenase;   ICDH = isocitric dehydrogenase;
β-OH-BuDH = β-hydroxybutyric dehydrogenase;   THFF = tetrahydrofolic acid formylase;
SDH = succinate dehydrogenase.

* Best system for analysis.

a Diaphorase used instead of phenazine methosulfate

b Tris buffer, no hydrazine added.

It was found that both D-α-hydroxy butyric and lactic acids are substrates for this enzyme. From 2-50 μg/ml of α-hydroxy butyrate can be determined in the presence of all acids (except lactic) (Table 5).

β-hydroxy butyrate dehydrogenase from <u>Rhodopseudomonas spheroides</u> is also a highly specific enzyme. Of all the acids tried (those listed in Table 5) Guilbault <u>et al</u>[76] found only D-β-hydroxy butyric and DL-citric acids reacted and were determinable. From 1-75 μg/ml concentrations of D-β-OH butyric acid was determined in the presence of glutamic, lactic, malic and α-hydroxybutyric acids with an accuracy of ± 1.3% and a precision of 2% (Table 6).

Thus a three component mixture of butyric, D-α-hydroxybutyric and D-β-hydroxybutyric acids can be analyzed using the 3 dehydrogenase enzyme systems: malate, lactate (Type IV) and β-hydroxy butyrate.

3.   DL-Citric and D-Isocitric Acids

Guilbault <u>et al</u>[76] found that both DL-citric and D-isocitric acids are substrates for the malate dehydrogenase, lactate dehydrogenase and glutamate dehydrogenase enzyme systems (Tables 4 and 5). More selectivity is achieved in the β-hydroxy butyrate and isocitrate dehydrogenase systems. Only β-hydroxybutyric acid interferes with the determination of DL-citric acid using β-hydroxy butyrate dehydrogenase. And only D-isocitric acid, of all the acids tried, reacted with isocitric dehydrogenase (Type IV, pig heart):

D-Isocitrate + NADP $\xrightarrow{\text{Dehydrogenase}}$

$$\alpha\text{-oxoglutarate} + CO_2 + NADPH + H^+$$

NADPH + Resazurin $\xrightarrow[\text{Sulfate}]{\text{Phenazine Methyl}}$ Resorufin + NADP

   (Non-Fluorescent)   (Fluorescent)

TABLE 6

Typical Analysis of Acids

D-β-Hydroxy Butyric Acid[a]			L-Glutamic Acid[b]			L-Lactic Acid[c]		
Added µg/ml	Found µg/ml	Rel Error %	Added µg/ml	Found µg/ml	Rel Error %	Added µg/ml	Found µg/ml	Rel Error %
1.00	1.01	+ 1.0	5.00	4.95	- 1.0	0.500	0.510	+ 2.0
5.00	4.90	- 2.0	15.0	14.7	- 2.0	1.50	1.47	- 2.0
10.0	10.0	0.0	25.0	25.0	0.0	3.00	3.08	+ 2.7
50.0	51.0	+ 2.0	75.0	77.0	+ 2.2	5.00	4.85	- 3.0
75.0	74.0	- 1.5	100.0	102.0	+ 2.0	10.0	10.0	0.0
Av. Rel. Error		± 1.3			± 1.4			± 1.9

[a] Analysis with β-hydroxy butyric dehydrogenase in the presence of 1 mg/ml of L-glutamic, L-lactic, L-malic, and D-α-hydroxy butyric acids.

[b] Analysis with glutamate dehydrogenase in the presence of 1 mg/ml of acetic, D-tartaric and D-β-hydroxy butyric acids.

[c] Analysis with lactate dehydrogenase Type IV in the presence of 1 mg/ml L-malic, L-glutamic, and D-β-hydroxy butyric acids.

The rate of production of the highly fluorescent resorufin is a measure of the D-isocitric acid present.

A four component mixture of DL-citric, D-isocitric, L-lactic and L-glutamic acids was analyzed[76] using the 4 enzymes β-hydroxy butyrate, isocitrate, lactate and glutamate dehydrogenases (Table 7). Excellent results were obtained.

Citrate can also be determined with citrase, an enzyme induced in E. coli[83,84] which catalyzes the breakdown of DL-citrate to oxaloacetate and acetate in the presence of an activator, Mg^{2+}:

$$\text{Citrate} \xrightleftharpoons{\text{Citrase}} \text{Acetate + oxaloacetate}$$

$$\text{Oxaloacetate} \xrightleftharpoons{\text{Decarboxylase}} CO_2 + \text{pyruvate}$$

An enzyme extract from A. aerogenes[85,86] catalyzes the breakdown of oxaloacetate to pyruvate and establishes a quantitative conversion of citrate by removal of the oxaloacetate formed. The pyruvate produced is monitored with the use of lactate dehydrogenase:

$$\text{Pyruvate} + \text{NADH} + H^+ \xrightleftharpoons{\text{LDH}} \text{lactate + NAD}$$

The rate of disappearance of NADH is proportional to the amount of citrate present. The A. aerogenes enzyme extract contains both citrase as well as oxaloacetate decarboxylase and can be used to effect the complete conversion of citrate to pyruvate. From 1-10 µg/ml of citrate are determinable; only L-glutamate and α-keto-glutarate interfere. The enzyme preparation is not available commercially.

Citrate can also be determined with aconitase[87], which catalyzes the conversion of citrate to isocitrate:

$$\text{Citrate} \xrightarrow{\text{Aconitase}} \text{Isocitrate}$$

TABLE 7

Analysis of a Mixture of Citric, Isocitric, Lactic and Glutamic Acids

Added, μg/ml				Found, μg/ml			
DL-Citric	D-Isocitric	L-Lactic	L-Glutamic	DL-Citric	D-Isocitric	L-Lactic	L-Glutamic
10.0	1.00	10.0	10.0	10.1	1.00	9.90	10.2
10.0	1.00	5.00	50.0	10.0	1.01	5.10	49.2
10.0	1.00	10.0	50.0	9.9	0.980	10.1	51.0
50.0	0.500	10.0	50.0	49.2	0.495	10.0	51.0

Rel. Error, %			
DL-Citric	D-Isocitric	L-Lactic	L-Glutamic
+ 1.0	0.0	- 1.0	+ 2.0
0.0	+ 1.0	+ 2.0	- 1.6
- 1.0	- 2.0	+ 1.0	+ 2.0
- 1.6	- 1.0	0.0	+ 2.0

The isocitrate is then determined with isocitrate dehydrogenase as above. The increase in absorbance of 340 or 366 mμ due to formation of NADPH is a measure of the reaction. Or resazurin can be used and the rate of production of the fluorescent resorufin measured.[76] Aconitase is also not available commercially, and this limits the ready adaptation of this procedure.

4. Formic Acid

Formic acid, in the 10-500 μg/ml concentration range, can be determined using malate dehydrogenase.[76] A prior separation of formate from interfering acids is necessary.

A specific method for formic acid is based on the use of tetrahydrofolic acid formylase[88] which catalyzes the following reaction:

Formate + ATP +

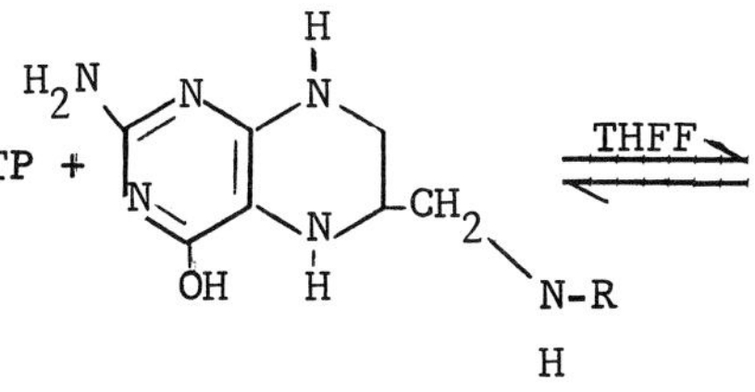

Tetrahydrofolic Acid

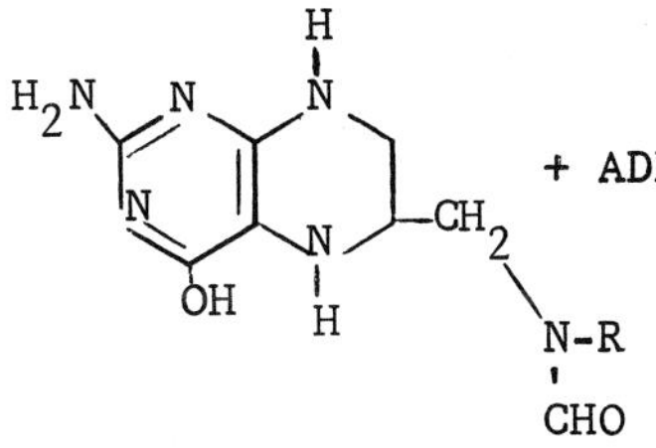

+ ADP + PO_4^{3-}

I

The reaction is monitored by conversion of product I to a highly colored species (II, λ_{max} = 350 mμ) with acid:

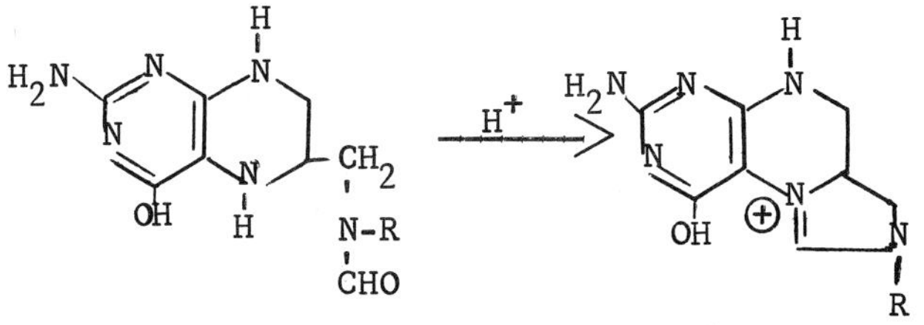

II

λ_{max} = 350 mμ

The enzyme is highly specific for formate; acetate, pyruvate and other common acids, aldehydes, amides and alcohols do not interfere. The enzyme is now available commercially (Koch-Light Labs, London) but the acid must be prepared. [88]

5.   Fumaric Acid

Fumaric acid can be determined by conversion to L-lactate with fumarase, malate dehydrogenase and lactate dehydrogenase [89,90]:

$$\text{Fumarate} + H_2O \xrightleftharpoons{\text{Fumarase}} \text{L-malate}$$

$$\text{L-malate} + \text{NAD} \xrightleftharpoons{\text{MDH}} \text{Pyruvate} + CO_2 + \text{NADH} + H^+$$

The CO_2 produced can be measured manometrically, or the NADH measured colorimetrically.

The fumarase is not available commercially, but must be prepared. [91]

6.   Hyaluronic Acid

The enzyme hyaluronidase is completely specific for hyaluronic acid; 50-1,000 µg/ml of substrate can be determined.

Assay procedures based on viscosity[92], salt-formation[93,94], and reducing sugar formation[95] have already been discussed (p. 63).

Alternatively, the enzymic reaction can be monitored by noting the increase in absorbance of the 4,5 unsaturated uronide formed at 230 mµ. Hyaluronic acid does not absorb at this wavelength.[96]

7.   L-Lactic Acid

a.   Determination With Lactate Dehydrogenase, Type II, and NAD. Type II lactate dehydrogenase (LDH) catalyzes the oxidation of L-lactate in the presence of the coenzyme NAD:

$$\text{L-lactate} + \text{NAD} \xrightleftharpoons{\text{LDH}} \text{pyruvate} + \text{NADH} + \text{H}^+$$

The equilibrium for this reaction lies far to the left, and a quantitative oxidation of L-lactate to pyruvate can be accomplished only 1) at high pH and 2) by adding a hydrazine to trap the α-keto acid formed through hydrazone formation:

$$\text{L-lactate} + \text{NAD} + \text{Hydrazine} \xrightleftharpoons{\text{pH 9-10}}$$

$$\text{pyruvate hydrazone} + \text{NADH}$$

The reaction can be monitored spectrophotometrically by the increase in absorbance due to the NADH produced. Or the reaction can be monitored fluorometrically by coupling with the resazurin-phenazine methosulfate indicator reaction.[76] The rate of production of the highly fluorescent resorufin is a measure of the lactate present.

In a thorough study of the specificity of lactate
dehydrogenase (Type II, rabbit muscle), Guilbault et
al[76] found that if hydrazine is added to react with
the α-keto acid formed and thus ensure a more complete
reaction from hydroxy to keto-acid, then the following
acids are oxidized, the order of decreasing reaction
rate being: glycolic ⟩L-lactic ⟩oxalic⟩ L-glutamic =
D-isocitric ⟩DL-citric⟩ L-malic (Table 4).　Acetic,
adipic, benzilic, butyric, D-α- and β-hydroxy butyric,
chloroacetic, formic, glutaric, L-malic, malonic,
phthalic, salicylic, DL-succinic, D-tartaric and
amino acids have no effect.　Using tris buffer, pH 9.5,

$$
\begin{array}{cccc}
\text{H} & \text{H} & \text{H} & \text{O} \\
| & | & | & || \\
CH_3\text{-}C\text{-}COOH & H\text{-}C\text{-}COOH & HOOC\text{-}CH_2\text{-}C\text{-}COOH & C\text{-}COOH \\
| & | & | & | \\
OH & OH & OH & OH
\end{array}
$$

　　Lactic　　　Glycolic　　　Malic　　　　　Oxalic

with hydrazine added, 4-175 μg/ml of L-lactic can
be determined.

　　b.　**Determination With Yeast Enzyme (Type IV)**.
Lactate dehydrogenase, Type IV from Bakers Yeast,
is an enzyme that does not require the use of NAD
or NADH.　Resazurin serves as an electron acceptor,
being reduced to the fluorescent resorufin.[76]
Using this enzyme from 0.2-10 μg/ml of L-lactate can
be determined specifically in the presence of all
acids except D-α-hydroxybutyric acid.　Greater sensitivity was found with the Type IV enzyme (as little
as 0.2 μg detectable compared to 4 μg with Type II).

　　Some results obtained in the determination of
lactic acid in the presence of 1 mg/ml concentrations of L-malic, L-glutamic and D-β-hydroxybutyric
acids are given in Table 6.　Table 7 lists the
results of the analysis of a four component mixture
of DL-citric, D-isocitric, L-lactic and L-glutamic
acids.

8.   L-Malic Acid

Malate Dehydrogenase (MDH) catalyzes the oxidation
of L-malic acid in the presence of the coenzyme NAD.
Like the glutamate and lactate dehydrogenase systems
the equilibrium of the malate dehydrogenase system lies
far to the left, and both high pH and hydrazine are
needed to effect a quantitative analysis of L-malic
acid.

The course of the reaction can be followed spectrophotometrically at 340 or 366 mμ, or fluorometrically
by using resazurin.[76] Guilbault et al[76] found
malate dehydrogenase to be a highly unspecific enzyme.
All the acids listed in Table 5 are substrates. Good
rates were obtained with all acids regardless of the
buffer used: tris, glycine, or glycine with hydrazine
added (Table 4).

Warburg[98] has proposed the use of the 3-acetyl-
pyridine analog of NAD (AP-NAD) without the use of a
trapping agent for oxaloacetate. The reaction is quantitative from left to right due to the favorable redox
potential of AP-NAD/AP-NADH in contrast to NAD/NADH.

$$\text{L-Malate + AP-NAD} \xrightarrow{\text{MDH}} \text{oxaloacetate + AP-NADH + H}^+$$

The AP-NADH is measured spectrophotometrically at 366 mμ.

9.   Succinic Acid

Succinic dehydrogenase (SDH) catalyzes the reaction:

$$\text{Succinate} \xrightarrow{\text{SDH}} \text{Fumarate + 2 H}^+$$

The reaction is monitored colorimetrically using ferricyanide as an electron acceptor for the enzyme:

$$\text{Succinate + 2 [Fe(CN)}_6\text{]}^{3-} \xrightarrow{\text{SDH}} \text{fumarate + 2 H}^+ +$$
$$2 \text{ [Fe(CN)}_6\text{]}^{4-}$$

The decrease in the absorbance of ferricyanide
at 450 mμ is measured and related to the concentration
of succinic acid present.[99]   From 5-100 μg/ml of
succinate can be determined specifically (Table 5).

F. HYDROXY COMPOUNDS, ESTERS AND ALDEHYDES
1.   Ethanol and Aliphatic Alcohols
    a.   Assay with Alcohol Dehydrogenase. Alcohol dehy-
drogenase (ADH) reversibly oxidizes ethanol and several
other alcohols to their corresponding aldehydes in the
presence of nicotinamide adenine dinucleotide (NAD):

$$CH_3-CH_2-OH + NAD \xrightarrow{ADH} CH_3CHO + NADH + H^+$$

The reaction may be monitored by following the absorbance
change due to production of NADH. Guilbault and Kramer
have described a fluorometric procedure for measuring
this reaction.[75]   Resazurin is used with phenazine
methosulfate or diaphorase. The rate of production of
the highly fluorescent resorufin is a measure of the
concentration of ethanol.[100]   As little as 0.10 μg is
determinable.

    In addition to ethanol, allyl alcohol, n-propanol,
n-butanol, isopropanol and n-amyl alcohol are substrates
of this enzyme and can be determined.[100]   Some reaction is also noted with glycerol, methanol, ethylene,
glycol and isobutanol.

    Mark[101] has described a kinetic method for the
analysis of mixtures of alcohols employing enzyme
catalyzed reactions. The method of proportional equations was modified and applied to the determination of
ethanol and n-propanol, both catalytically oxidized by
alcohol dehydrogenase but at different rates.

    b.   Assay with Alcohol Oxidase. In recent publica-
tions Janssen and co-workers[102,103] have described
the isolation of a novel enzyme, designated "alcohol

oxidase," from the mycelium of a Basidiomycete belonging
to the Polyporoceae family. In the presence of O_2, this
enzyme catalyzes the oxidation of the lower primary
alcohols to the corresponding aldehydes and H_2O_2. It
was reported that unsaturated alcohols are also good
substrates, but that branched chain and secondary alcohols
are not attacked. The authors[102,103] used a colorimetric peroxidase-o-dianisidine reagent to assay the
enzymic activity:

$$\text{Alcohol} \xrightarrow{\text{Oxidase}} H_2O_2$$

$$H_2O_2 + \text{o-dianisidine} \xrightarrow{\text{Peroxidase}} \text{"Colored Product"}$$

Guilbault and Sadar[100] used this enzyme to develop
fluorometric methods for the assay of 0.10-10 µg/ml
concentrations of methanol, ethanol, propanol, butanol
and allyl alcohol which are substrates of the enzyme.
The rate of production of the fluorescent oxidized p-
hydroxyphenylacetic acid (VI) is proportional to the
concentration of alcohol present. This enzyme appears
to be more selective in its activity than alcohol dehydrogenase and thus offers advantages over the latter.
The following alcohols and hydroxy compounds are not
oxidized at any appreciable rate: isobutanol, sec-
butanol, isopropanol, benzyl alcohol, anisyl alcohol,
glycerol, ethylene glycol, glycolic acid, α-hydroxy
butyric acid, lactic acid, cyclohexanol, phenol, prop-
anediols, butanediols, ethanolamine and methyl cellosolve.[100,102,103] Some Michaelis constants obtained
by Janssen and co-workers using the colorimetric monitoring system[102,103] and by Guilbault and Sadar using
the fluorometric indicator reaction[100] are listed
in Table 8. The lower K_m values obtained by the latter
workers reflect the greater sensitivity of the fluorescence method.

TABLE 8

Michaelis Constants for Alcohol Oxidase

| Alcohol | Km, mM | |
	Janssen et al	Guilbault and Sadar
Ethyl alcohol	10.0	1.82
Methyl alcohol	1.52	0.42
Butyl alcohol	133.	9.8
Propyl alcohol	54.6	19.0
Allyl alcohol	--	6.45

2.   Glycerol and Dihydroxy Acetone

a.   <u>Determination with Glycerol Dehydrogenase.</u> Glycerol dehydrogenase (GDH) catalyzes the oxidation of glycerol to dihydroxy acetone in the presence of NAD:

$$\text{Glycerol} + \text{NAD} \xrightleftharpoons{\text{GDH}} \text{Dihydroxyacetone} + \text{NADH} + \text{H}^+$$

The reaction can be monitored by the formation of NADH spectrophotometrically at 340 or 366 mμ.

The dehydrogenation is complete at pH 10-11.   At lower pH (pH 6-8) the reduction of dihydroxy acetone to glycerol is quantitative, and this reaction can be used for the assay of dihydroxy acetone.   The decrease in absorbance at 340 or 366 mμ, due to oxidation of NADH is observed and is proportional to the concentration of this compound.

Guilbault and Sadar[100] have described a fluoro-metric method for the assay of glycerol.   The non-fluorescent dye resazurin is used, and the rate of production of the fluorescent resorufin is proportional to the amount of glycerol present.   Frings and Pardue[104] have described a coupled colorimetric readout reaction for use in an automatic spectrophotometric method for the enzymatic determination of glycerol.   The NADH formed reacts with a dye, 2,6-dichloroindophenol:

NADH + H$^+$ + Oxidized dye $\xrightarrow{\text{diaphorase}}$ NAD + reduced dye

(Blue)   (Colorless)

The rate of disappearance of the blue colored species
is measured at 600 mμ.

    b. <u>Determination of Glycerol with Glycerokinase (GK)</u>.
Glycerol is specifically phosphorylated with glycerokinase
(GK) in the presence of adenosine triphosphate, ATP,
and an activator (Mg^{++}) to give L-glycerol-1-phosphate:

Glycerol + ATP $\xrightarrow[\text{Mg}^{2+}]{\text{GK}}$ L-glycerol-1-phosphate + ADP

The L-glycerol-1-phosphate formed is oxidized with
glycerol phosphate dehydrogenase (GPDH) in the presence
of the coenzyme, NAD:

L-Glycerol-1-phosphate + NAD $\xrightarrow{\text{GPDH}}$

NADH + dihydroxy
acetone phosphate + H$^+$

The amount of NADH produced, as measured spectrophotometrically,is proportional to the amount of glycerol
present. The equilibrium of the GPDH reaction is displaced to the right by the addition of hydrazine and use
of a high pH (9 or 10).

    This reaction can also be monitored fluorometrically
by use of the resazurin indicator reaction described
by Guilbault and Kramer.[75] The NADH produced in the
GPDH reaction effects the reduction of resazurin to
resorufin. The rate of production of the fluorescence,
ΔF/min, is proportional to the glycerol present.

    The reaction sequence described is completely specific
for glycerol, since the GPDH enzyme has only one substrate, L-glycerol-1-phosphate.

The glycerokinase procedure for glycerol, though more specific than the glycerol dehydrogenase procedure, also involves the use of more costly reagents. If other substrates that are interferences in the GDH procedure are known to be absent then the glycerol dehydrogenase method is the method of choice.[100]

3.   Hydroxy Pyruvate

Hydroxy pyruvate can be determined by an enzyme catalyzed reduction to glycerate in the presence of NADH. Lactate dehydrogenase from animal tissue[105] effects the reduction to L-glycerate, whereas an enzyme from plant tissues[105,106] catalyzes a reduction to D-glycerate.

Hydroxy pyruvate + NADH + H^+ ———> Glycerate + NAD

The reaction can be monitored by the decrease in absorbance at 366 mμ. Glyoxylate is the only interference.

4.   Butylene Glycol

Butylene glycol dehydrogenase from <u>Aerobacter aerogens</u> or <u>staphylococcus aureus</u>[107] catalyzes the dehydrogenation of butylene glycol in the presence of NAD:

$$CH_3\text{-}\underset{\underset{OH}{|}}{CH}\text{-}\underset{\underset{OH}{|}}{CH}\text{-}CH_3 + NAD \longrightarrow CH_3\text{-}COCHOHCH_3 + NADH$$

The reaction is followed spectrophotometrically at 340 mμ by NADH production.

Propane 1,2-diol is also a substrate and can be determined.[108] The following do not interfere: ethanol, ethylene glycol, sorbitol, lactic and malic acids, glycerol, tartaric acid, isopropanol, isoamyl alcohol and glucose.

5.   Sorbitol

Sorbitol dehydrogenase (SoDH), an enzyme occurring in mammalian liver and rat kidney[109], catalyzes the specific dehydrogenation of sorbitol:

$$\text{Sorbitol} + \text{NAD} + O_2 \underset{\longleftarrow}{\overset{\text{SoDH}}{\longrightarrow}} \text{D-fructose} + \text{NADH} + H^+$$

The reaction can be monitored manometrically be determination of oxygen uptake with methylene blue as H-acceptor[109] or spectrophotometrically by the formation of NADH.

The enzyme catalyzes the oxidation of sorbitol, L-iditol, allitol, xylitol, ribitol, and some heptitols, but not mannitol, erythritol, arabitol and D-iditol.[110,111]

6.   Triglycerides

Triglycerides are determined by hydrolysis, either enzymic or non-enzymic, to glycerol. The glycerol produced is then determined as mentioned previously.

$$\text{Triglyceride} \longrightarrow \text{Glycerol}$$

Kaufman and Wessels[112] suggested the use of chromatography and enzyme hydrolysis for the selective determination of triglyceride structure, and Eggstein and Kreutz[113] and Spinella and Mager[114] used a modified enzymatic method for the assay of blood and plasma neutral fats.

Vela et al[115,116] has suggested the use of pancreatic lipase for the selective hydrolysis of glycerides in the detection of esterified oils in olive oil.

7.   Phenols

A phenol oxidase, prepared from cultured mushrooms and dandelion roots, was described by Drawert, Gebbing and Ziegler[117] for the detection of phenols on thin layer plates. Procedures were developed for the determination of pyrocatechol, pyrogallol, 3,4-dihydroxy phenylacetic acid, caffeic acid, gallic acid, catechol and other phenols.

8.   Xanthine and Hypoxanthine

Many methods have been proposed for the assay of
xanthine and hypoxanthine using xanthine oxidase (pp.89-
92).   Both hypoxanthine and xanthine can be determined
if spectrophotometric measurements are made at two wavelengths[118,119] (Fig. 4).   Hypoxanthine exhibits an
absorbance maximum at 250 mμ, xanthine at 270 mμ and
uric acid at 293 mμ.   The decrease in absorbance at
250 mμ after addition of xanthine oxidase is proportional
to the hypoxanthine present.   The decrease in absorbance
at 293 mμ after addition of uricase is a measure of the
sum of hypoxanthine and xanthine.

$$\text{Uric Acid} + O_2 + 2\ H_2O \xrightarrow{\underline{Uricase}} \text{allantoin} + H_2O_2 + CO_2$$

Guilbault et al [120]   described an electrochemical
method for the determination of hypoxanthine and xanthine
(pp. 90-91).   The slopes of the depolarization curves,
ΔE/min, were found to be proportional to the concentration of hypoxanthine and xanthine.

Guilbault, Brignac and Zimmer[4] have described a
fluorometric method for the assay of nanogram quantities
of xanthine and hypoxanthine in biological samples using
the homovanillic acid-peroxidase indicator reaction
(pp. 90, 92).   From 0.0030 to 1.00 μg/ml of hypoxanthine
and xanthine were analyzed.

Xanthine oxidase catalyzes the oxidation of at least
30 aldehydes, ketones and purines.[121,122]   Talsky and
Fink[123] have used xanthine oxidase to determine a
number of aldehydes.

9.   Steroid Alcohols

A specific method for the determination of urinary
steroid alcohols involves their catalytic oxidation
with a hydroxysteroid dehydrogenase (HSD).   Several
HSD s have been purified and can be listed in terms

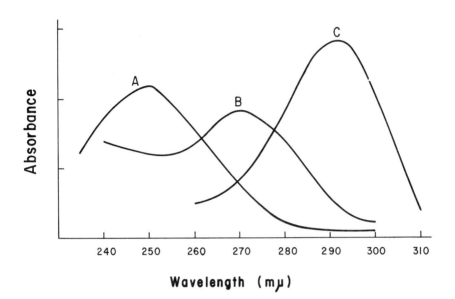

FIG. 4

Spectra of equimolar concentrations of
hypoxanthine (A), xanthine (B) and
uric acid (C).

of substrate and pyridine nucleotide specificity:

a. 3-α-HSD interconverts 3-α-hydroxy and 3-keto steroids:

$$3\text{-}\alpha\text{-Hydroxysteroid} + NAD \xrightarrow{\text{HSD}} 3\text{-ketosteroid} + NADH$$

The enzyme is purified from <u>Pseudomonas testosteroni</u>[124], or rat liver.[125] The latter works equally well with NADP.

b. 3-β, 17-β-HSD converts 3-β and 17-β-hydroxy-steroids to the corresponding 3- and 17-keto steroids. It likewise is obtained from <u>Pseudomonas testosteroni</u>[126] by the addition of testosterone.

$$3\text{-}\beta\text{-Hydroxysteroid} + NAD \rightleftharpoons 3\text{-ketosteroid} + NADH$$

$$17\text{-}\beta\text{-Hydroxysteroid} + NAD \rightleftharpoons 17\text{-ketosteroid} + NADH$$

c. An 11-β-HSD from liver microsomes interconverts cortisol and cortisone in the presence of NAD or NADP[127]

$$Cortisol + NAD \xrightarrow{11\text{-}\beta\text{-HSD}} Cortisone + NADH$$
$$(NADP) \hspace{4cm} (NADPH)$$

d. A steroid induced 20-β-HSD from Streptomyces[128] promotes the oxidation of 20-β-hydroxysteroids to 20-ketosteroids:

$$20\text{-}\beta\text{-Hydroxysteroid} + NAD \xrightleftharpoons{20\text{-}\beta\text{-HSD}} 20\text{-ketosteroid} + NADH$$

All of these hydroxysteroid dehydrogenases can be used to assay for the corresponding hydroxysteroid by spectro-photometric measurement of the NADH produced or for the assay of the ketosteroid by measurement of the decrease of NADH at 340 mμ. Generally, hydrazine is added to trap the ketosteroids already present in urine and those formed by the action of the enzymes in an assay of the hydroxysteroid. The enzymes are fairly specific for their corresponding hydroxysteroids.

10. Detection and Determination of Esters

West and Qureshi have proposed a qualitative test
for esters using methyl red and lipase.[129] A change
in the color of the solution from yellow to red caused
by the lipolytic hydrolysis of the ester indicated the
presence of the compound. The test was found to be more
specific than the hydroxamic acid test[129]

$$\text{Ester + Methyl Orange} \xrightarrow{\text{Lipase}} \text{Acid + Methyl Orange}$$

    (yellow)   (red)

Of the lipases tested steapsin gave the best results.
A number of non-esters, which give a positive hydroxamic acid test, do not give an enzymic test: lactones,
anhydrides, formic and phthalic acids, acetamide, chloral
hydrate, and chloroform.

Guilbault, Kramer and Cannon[130] used an electrochemical method for the analysis of thiocholine esters.
The rate of depolarization of a platinum electrode
$\Delta E/\Delta t$, was proportional to the amount of thioester
present.

Billiar and Eik-Nes[131] used cholinesterase to
determine steroid acetates. A steroid dehydrogenase
could be used to determine the steroid liberated.

G. INORGANIC SUBSTANCES

1. Peroxide

Colorimetric, electrochemical and fluorometric methods
for the assay of peroxide using the enzyme peroxidase
have been already discussed in Chapter 3 (pp. 87-89).

Weetall and Weliky[132] described an enzyme-
impregnated paper, prepared by chemically coupling
horse radish peroxidase to carboxymethyl cellulose paper
strips in the presence of N,N-dicyclohexylcarbodiimide,

for the detection of small amounts of peroxide. The
immobilized peroxidase was used for the colorimetric
assay of peroxide in the same way the soluble enzyme
is used (p. 87).

Peroxidase is specific for inorganic peroxides.
Although some organic peroxides are substrates, the
analytical results are non-reproducible.

2. Ammonia

Since ammonia is a substrate for glutamate dehydrogenase, this enzyme system can be used for its
specific assay:

$$NH_4^+ + H^+ + NADH + \alpha\text{-ketoglutarate} \underset{\longleftarrow}{\overset{GDH}{\longrightarrow}}$$

$$glutamate + NAD + H_2O$$

This reaction can be monitored by the decrease in
absorbance at 340 mμ. The enzymatic procedure is
specific for ammonia in the presence of amines.

Faway and Dahl[133] used this procedure for the
enzymatic determination of ammonia in tissue body
fluids; Mondzac, Ehrlich and Seegmiller[134] for its
assay in body fluids; Kirsten, Gerez and Kirsten[135]
for its determination in blood; and Reichelt, Kjamme
and Tveit[136] for analysis in blood and tissue.

Roch-Ramel[63] increased the sensitivity of the
assay by measuring the decrease in fluorescence of
NADH (λ_{ex} = 340 mμ; λ_{em} = 460 mμ). From 4×10^{-11}
to 2×10^{-10} equivalents of NH_4^+ were determinable.

3. Nitrate

Nitrate reductase, found in fungi[137] and
higher plants, catalyzes the reduction of nitrate
to nitrite:

$$NO_3^- + H^+ + NADH \longrightarrow NO_2^- + H_2O + NAD$$

The reaction can be monitored spectrophotometrically

by the disappearance of NADH, or by colorimetric measurement of the nitrite formed based on the production of a
red colored azo dye (the Griess-Ilosvay reaction.)[138]

This method is completely specific for nitrate in the
absence of nitrite (if the nitrite produced is measured
by the Griess-Ilosvay reaction). Chlorate is also reduced
by nitrate reductase.[139]

4.   Hydroxylamine

Hydroxylamine reductase, an enzyme from Neurospora[140]
catalyzes the reduction of hydroxylamine to ammonia:

$$NH_2OH + NADH + H^+ \longrightarrow NH_3 + NAD + H_2O$$

The enzyme is specific for hydroxylamine. The reaction
can be followed spectrophotometrically by noting the
decrease in absorbance of NADH at 340 mμ; or by colorimetric measurement of the remaining hydroxylamine by
iodine oxidation to nitrite, which is then assayed by
the Griess-Ilosvay reaction as described above.[141]

5.   Carbon Dioxide

The enzyme carbonic anhydrase (CA) can be used for
the specific detection and determination of carbon
dioxide, in products such as carbonated wine, soft
drinks, etc.[142-144] The enzyme catalyzes the hydration

$$CO_2 + H_2O \xrightarrow{CA} H_2CO_3$$

of CO_2 to carbonic acid. The reaction can be monitored
by the pH change that occurs either potentiometrically or by the change in color of the pH indicator.

6.   Phosphate

Schulz, Passonneau and Lowry[145] nave described a
fluorometric enzymic method for the measurement of
inorganic phosphate based on the following sequence:

Glycogen + phosphate $\xrightarrow{\text{phosphorylase a}}$ glucose-1-phosphate

Glucose-6-Phosphate $\xleftarrow{\boxed{\text{Phosphoglucomutase}}}$

NADP $\xrightarrow{\boxed{\text{Dehydrogenase}}}$ NADPH + 6-Phosphoglucono-

lactone + H^+

Phosphorylase a catalyzes the phosphorylation of glycogen to glucose-1-phosphate. This is then converted to glucose-6-phosphate by phosphoglucomutase. The glucose-6-phosphate produced is detected in an indicator reaction with glucose-6-phosphate dehydrogenase and NADP. The NADPH produced is fluorescent and the rate of its formation indicates the phosphate present at concentrations of 2×10^{-12} to 5×10^{-10} moles.

Faway, Roth and Faway [146] assayed inorganic phosphate in tissue and serum using this same reaction sequence, except for a spectrophotometric measurement of NADPH at 340 mμ.

7.   Pyrophosphate

Inorganic pyrophosphatase [147] catalyzes the conversion of pyrophosphate to orthophosphate:

$$P_2O_7^{4-} + H_2O \xrightarrow{\text{pyrophosphatase}} 2\ HPO_4^{2-}$$

The orthophosphate formed is determined colorimetrically as the phosphomolybdate ($\lambda_{max} = 600$ mμ).

The reaction is completely specific for pyrophosphate; there are no interferences. As little as one microgram of pyrophosphate is determinable.

REFERENCES

1. E. F. Gale, Biochem, J. <u>39</u>, 46 (1945).

2. B. L. Horecker and P. Smyrniotis, Biochem. Biophys Acta <u>12</u>, 98 (1953).

3.   S. Kaufman, J. Biol. Chem. 216, 153 (1955).

4.   G. G. Guilbault, P. Brignac and M. Zimmer,
     Anal. Chem. 40, 190 (1968).

5.   S. P. Colowick and N. O. Kaplan, eds., Methods
     of Enzymology, Academic Press, New York, 1957.

6.   D. Keilin and E. F. Hartree, Biochem. J. 42,
     230 (1948).

7.   G. G. Guilbault, B. Tyson, P. Cannon and D. N.
     Kramer, Anal. Chem. 35, 582 (1963).

8.   G. Charlton, D. Read and J. Read, J. Appl.
     Physiol. 18, 1247 (1963).

9.   A. H. Kadish and D. A. Hall, Clin. Chem. 9,
     869 (1965).

10. Y. Makino and K. Koono, Rinsho Byori 15, 391 (1967).

11. A. Kadish, R. Litle and J. Sternberg, Clin. Chem.
     14, 116 (1968).

12. S. J. Updike and G. P. Hicks, Nature 214, 986 (1967).

13. A. Martinsson, J. Chromatog. 24, 487 (1966).

14. E. Kawerau, Z. Klin. Chem. 4, 224 (1966).

15. Fyowa Fermentation Industry, French Patent
     1,410,747 (1967).

16. R. E. Phillips and F. R. Elevitch, Am. J. Clin.
     Pathol., in press.

17. G. G. Guilbault, Anal. Chem. 36, 529R (1966).

18. G. G. Guilbault, Anal. Chem. 40, 459R (1968).

19. B. Naganna, M. Rajamma and K. V. Rao, Clin. Chim.
     Acta 17, 219 (1967).

20. L. Hollister, E. Helmke and A. Wright, Diabetes
     15, 691 (1966).

21. H. V. Malmstadt and H. L. Pardue, Anal. Chem.
     33, 1040 (1961).

22. Ibid., Clin. Chem. 8, 606 (1962).

23. H. Pardue, R. Simon and H. Malmstadt, Anal.
     Chem. 36, 735 (1964).

24. H. Pardue, Anal. Chem. 35, 1240 (1963).

25. W. J. Blaedel and C. Olson, Anal. Chem. 36,
     34 (1964).

26. G. G. Guilbault and P. Hodapp, Anal. Letters
     1, 789 (1968).

27. H. U. Bergmeyer and H. Moellering, Clin. Chim.
     Acta 14, 74 (1966).

28. C. F. Boehringer and G. Soehne, Neth. Patent
    Appl. 6,611,550 (Feb., 1967).

29. G. G. Guilbault, P. Brignac and M. Juneau, Anal.
    Chem. 40, 1256 (1968).

30. G. Avigad, D. Amaral, C. Asensino and B. Horecker,
    J. Biol. Chem. 237, 2742 (1962).

31. M. R. McDonald in Methods in Enzymology, (S.P.
    Colowick and N. O. Kaplan, eds.), Academic Press,
    New York, 1955, Vol. 1, p. 326.

32. G. G. Guilbault, M. Sadar, and K. Peres, Anal.
    Chim. Acta, in press.

33. G. Pfleiderer and L. Grein, Biochem. Z. 328,
    499 (1957).

34. C. Rerup and I. Lundquist, Acta Pharmocol. Toxicol.
    25, 41 (1967).

35. J. V. Passonneau, P. D. Gatfield, D. W. Schulz and
    O. H. Lowry, Anal. Biochem 19, 315 (1967).

36. A. Hu, R. Wolfe and F. Reithel, Arch. Biochem.
    Biophys. 81, 500 (1959).

37. R. D. deMoss, in Methods in Enzymology, (S. P.
    Colowick and N. O. Kaplan, eds.), Academic Press,
    New York, 1957, Vol. III, p. 232.

38. I. G. Leder, J. Biol. Chem. 225, 125 (1957),

39. E. A. Zeller in The Enzymes. Vol. 8, (P. Boyer,
    H. Lardy and K. Myrback, eds.), Academic Press,
    New York, 1963, pp. 313-335.

40. J. R. Fouts, L. A. Blanksma, J. A. Carlon and
    E. A. Zeller, J. Biol. Chem. 225, 1025 (1957).

41. E. A. Zeller, J. R. Fouts, J. A. Carlon, J. C.
    Lazanas and W. Voegtti, Helv. Chim. Acta 39, 1632
    (1956).

42. P. A. Shore, A. Burkholter and V. H. Cohn, J.
    Pharmacol. Exptl. Therap. 127, 182 (1959).

43. V. H. Cohn and P. A. Shore, Anal. Biochem. 2, 237
    (1961).

44. B. Holmstedt and R. Tham, Acta Physiol. Scand. 45,
    152 (1959).

45. B. Holmstedt, L. Larsson and T. Tham, Biochem et
    Biophys. Acta 48, 182 (1961).

46. E. A. Zeller, Helv. Chim. Acta 23, 1509 (1940)

47. B. M. Braganca, J. H. Quastel and R. Schucher,
    Arch. Biochem. Biophys. 52, 18 (1954).

48. G. G. Guilbault, J. Montalvo and R. Smith, Anal.
     Chem., in press.

49. H. Blaschko in The Enzymes, (P. Boyer, H. Lardy and
     K. Myerback, eds.), Vol. 8, Academic Press, New York,
     1963, pp. 337-351.

50. G. G. Guilbault and P. Brignac, Anal. Chem., in
     press.

51. V. Bachrach and B. Reches, Anal. Biochem. 17, 38
     (1966).

52. C. McEqen and A. Sober, J. Biol. Chem. 242, 3068
     (1967).

53. G. Werner and N. Seiler, Drug. Res. Wuerttemberg
     15, 189 (1965).

54. O. Folin and H. Wu, J. Biol. Chem. 38, 81 (1919).

55. L. D. Scott, Brit. J. Exp. Pathol. 21, 93 (1940).

56. E. J. Conway, Biochem. J. 27, 419, 430 (1933).

57. L. Naftalm, J. F. Whitaker and A. Stephens,
     Clin. Chim. Acta 14 , 771 (1966).

58. R. W. Wilson, Clin. Chem. 12, 360 (1966).

59. M. Cirje and D. Sandru, Viata Med. 12, 1617 (1965).

60. A. Parmense, Sez. 1, 37 (5), 557 (1966).

61. C. Manzini, Minerva Med. 57, 385 (1966).

62. H. Kaltwasser and H. Schlegel, Anal. Biochem. 16,
     132 (1966).

63. F. Roch-Ramel, Anal. Biochem. 21 372 (1967).

64. I. Nielsen, Scand. J. Clin. Lab. Invest. 14,
     513 (1962).

65. H. V. Malmstadt and E. Piepmeier, Anal. Chem. 37,
     34 (1965).

66. W. C. Purdy, G. D. Christian and E. C. Knoblock,
     Presented at Northeast Section, Am. Assoc. of
     Clin. Chemists, 16th National Meeting, Boston,
     Mass., August, 1964.

67. S. A. Katz, Anal. Chem. 36, 2500 (1964).

68. S. A. Katz and G. A. Rechnitz, Z. Anal. Chem. 196,
     (4), 248 (1963).

69. H. A. Krebs in The Enzymes, Vol. 2, Academic Press,
     New York, 1951, p. 508.

70. T. Wieland, Angew Chem. 60, 171 (1951).

71. G. G. Guilbault and J. Hieserman, Anal. Biochem.,
     26, 1 (1968).

72. E. F. Gale, Adv. in Enzymology 6, 1 (1946).

73. E. F. Gale, Methods of Biochemical Analysis,
     Vol. 4, (D. Glick, ed.,) Interscience, New York,
     1957, p.285.

74. R. W. McGilvery and P. P. Cohen, J. Biol. Chem.
     174, 813 (1948).

75. G. G. Guilbault and D. N. Kramer, Anal. Chem.
     37, 1219 (1965).

76. G. G. Guilbault, R. McQueen and S. Sadar, Anal.
     Chim. Acta, in press.

77. J. M. Siegel, G. A. Martgomery and P. M. Bock,
     Arch. Biochem. Biophys. 82, 288 (1959).

78. J. R. Stern, S. Ochoa and F. Lynen, J. Biol. Chem.
     198, 313 (1952).

79. G. Pfleiderer, W. Gruber and T. Wieland, Biochem.
     Z. 326, 446 (1955).

80. M. L. Tanzer and C. Gilvarg, J. Biol. Chem. 234,
     3201 (1959).

81. F. Lundquist, U. Fungmann and H. Rasmussen,
     Biochem. J. 80, 393 (1961).

82. A. C. Bratton and E. K. Marshall, J. Biol. Chem.
     128, 537 (1939).

83. M. Grunberg-Manago and I. C. Gunsalus, Bact. Proc.
     73 (1953).

84. S. Dagley, J. Gen. Microbiol. 11, 218 (1954).

85. S. Dagley and E. A. Dawes, Nature 172, 345 (1953).

86. Ibid., Biochim. Biophys. Acta 17, 177 (1955).

87. S. R. Dickman and A. A. Cloutier, J. Biol. Chem.
     188, 379 (1951).

88. J. C. Rabinowitz and W. E. Pricer, J. Biol. Chem.
     229, 321 (1957).

89. M. L. Blanchard, S. Korkes, A. del Campillo and
     S. Ochoa, J. Biol. Chem. 187, 875 (1950).

90. S. Korkes, A. del Campillo and S. Ochoa, op. cit.
     p. 891.

91. V. Massey in Methods of Enzymology, (S. Colowick
     and N. Kaplan, eds.), Academic Press, New York,
     1955, Vol. 1, p. 729.

92. G. I. Swyer and C. W. Emmens, Biochem J. 41, 29
     (1947).

93. N. Ferrante, J. Biol. Chem. 220, 303 (1956).

94. S. Tolksdorf, M. McCready, D. McCullagh and
     E. Schwenk, J. Lab. Clin. Med. 34, 74 (1949).

95. M. Rapport, K. Meyer and A. Linker, J. Biol.
     Chem. 186, 615 (1950).

96. H. Greiling, Z. Physiol. Chem. 309, 239 (1957).

97. G. G. Guilbault, D. N. Kramer and E. Hackley,
     Anal. Biochem. 18, 241 (1967).

98. O. Warburg, Wasserstoffubertragende Fermente.
     Verlag. Berlin, 1948.

99. T. P. Singer, E. B. Kearney and P. Bernath,
     J. Biol. Chem. 223, 599 (1956).

100. G. G. Guilbault and S. H. Sadar, Anal. Chem.,
     in press.

101. H. Mark, Anal. Chem. 36, 1668 (1964).

102. F. W. Janssen, R. Kerwin and H. Ruelius,
     Biochem. Biophys. Res. Comm. 20, 530 (1965).

103. F. W. Janssen and H. Ruelius, Biochim. Biophys.
     Acta 151, 330 (1968).

104. C. Frings and H. Pardue, Anal. Chim. Acta 34,
     225 (1966).

105. H. E. Stafford, A. Magaldi and B. Vennesland,
     J. Biol. Chem. 207, 621 (1954).

106. H. Holzer and A. Holldorf, Biochem. Z. 329, 292
     (1957).

107. F. C. Happold and C. P. Spencer, Biochim. Biophys.
     Acta 8, 18, 543 (1952).

108. J. P. Gaubert and R. Gavard, Ann. Inst. Pasteur
     84, 734 (1953).

109. R. L. Blakley, Biochem. J. 19, 257 (1951).

110. J. McCorkindale and N. L. Edson, Biochem. J. 57,
     518 (1954).

111. H. G. Williams-Ashman, J. Banks and S. K. Wolfson,
     Arch. Biochem. Biophys. 72, 485 (1957).

112. H. P. Kaufman and H. Wessels, Fette Seifen
     Anstrichmittel 66, 13 (1964).

113. M. Eggstein and F. H. Kreutz, Ergeb Laboratoriumsmed
     2, 99 (1965).

114. C. J. Spinella and M. Mager, J. Lipid Res. 7,
     167 (1966).

115. F. M. Vela, Grasas Aeites Seville, Spain 16,
     69 (1965); Anal. Abstr. 13, 3230 (1966).

116. F. M. Vela, A. V. Roncero, F. R. Ayerbe and J. M. Moreno, op. cit. 15, 12 (1964); Anal. Abstr. 12, 2515 (1965).

117. F. Drawert, H. Gebbing and A. Ziegler, J. Chromatogr. 30, 259 (1964).

118. H. M. Kalckar, J. Biol. Chem. 167, 429 (1947).

119. P. Plesner and H. M. Kalckar, Methods of Biochemical Analysis, Vol. 3, (D. Glick, ed.), Interscience, New York, 1956, p. 103.

120. G. G. Guilbault, D. N. Kramer and P. L. Cannon, Anal. Chem. 36, 606 (1964).

121. H. J. Coombs, Biochem. J. 21, 1259 (1927).

122. V. H. Booth, Biochem. J. 32, 494 (1938).

123. G. Talsky and G. Fink, Z. Physiol. Chem. 348, 114 (1967).

124. P. Talalay and P. I. Marcus, J. Biol. Chem. 218, 675 (1956).

125. A. Tomkins, J. Biol. Chem. 218, 437 (1955).

126. P. Talalay, M. M. Dobson and D. F. Tapley, Nature 170, 620 (1952).

127. B. Hurlock and P. Talalay, Arch. Biochem. Biophys. 80, 468 (1959).

128. H. J. Hübener, F. Sahrholz, J. Schmidt-Thome, G. Nesemann and R. Junk, Biochim. Biophys. Acta 35, 270 (1959).

129. P. W. West and M. Qureshi, Anal. Chim. Acta 26, 506 (1962).

130. G. G. Guilbault, D. N. Kramer and P. Cannon, Anal. Chem. 34, 842 (1962).

131. R. B. Billiar and K. B. Eik-Nes, Anal. Biochem. 13, 11 (1965).

132. H. Weetall and N. Weliky, Anal. Biochem. 14, 160 (1966).

133. G. Faway and K. V. Dahl, Lebanese Med. J. 16, 169 (1963).

134. A. Mondzac, G. Ehrlich and J. E. Seegmiller, J. Lab. Clin. Med. 66, 526 (1965).

135. E. Kirsten, C. Gerez and R. Kirsten, Biochem. Z. 337, 312 (1963).

136. K. L. Reichelt, E. Kjamme and B. Tveit, Scand. J. Clin. Lab. Invest. 16, 433 (1964).

137. F. Egami and R. Sato, J. Chem. Soc. Japan $\underline{68}$,
      39 (1947).

138. D. J. Nicholas and A. Nason, Methods in Enzy-
      mology $\underline{3}$, 981 (1957).

139. R. M. Hill, H. Pivirick, W. E. Engelhard and
      M. Bogard, Agr. Food Chem. $\underline{7}$, 291 (1959).

140. M. Zucker and A. Nason, J. Biol. Chem. $\underline{213}$,
      463 (1955).

141. T. Z. Czaky, Acta Chem. Scand. $\underline{2}$, 450 (1948).

142. R. L. Morrison, J. Assoc. Off. Agr. Chemists
      $\underline{45}$, 627 (1962).

143. R. L. Morrison, J. Assoc. Off. Agr. Chemists
      $\underline{46}$, 288 (1963).

144. R. L. Morrison, J. Assoc. Off. Agr. Chemists
      $\underline{47}$, 711 (1964).

145. D. W. Schulz, J. V. Passonneau and O. H. Lowry,
      Anal. Biochem. $\underline{19}$, 300 (1967).

146. E. Faway, L. Roth and G. Faway, Biochem. Z.
      $\underline{344}$, 212 (1966).

147. K. Bailey and E. C. Webb, Biochem. J. $\underline{38}$, 394
      (1944).

CHAPTER 5

DETERMINATION OF ACTIVATORS
AND COENZYMES

A. GENERAL

An enzyme activator converts an inactive enzyme or an enzyme with low activity into an active biological catalyst, generally at very low concentrations:

$$E(\text{inactive}) + \text{Activator} \rightleftharpoons E(\text{active})$$

The activity of the enzyme will increase until enough activator has been used to activate the enzyme fully. The initial rate of the enzyme reaction is proportional to the activator concentration at low concentrations of activator.

Coenzymes are essential reactants and are consumed in the reaction. Analytical methods for coenzymes will be discussed in this chapter also.

B. DETERMINATION OF INORGANIC SUBSTANCES

1. Magnesium, Manganese and Zinc

A method for magnesium in plasma was described by Baum and Czok[1] based on the activation of isocitric dehydrogenase (ICDH):

$$\text{Isocitrate} + \text{NADP} \xrightarrow[\text{Mg}^{++}]{\text{ICDH}} \text{NADPH} + \text{H}^+ + \alpha\text{-oxoglutarate} \quad (1)$$

Without Mg^{++} ICDH shows no activity, and isocitrate is not oxidized. Using constant amounts of ICDH and non-rate-limiting concentrations of isocitrate and NADP, the rate of the enzymic reaction is proportional to magnesium in the concentration range $10^{-6}\underline{M}$ to $2 \times 10^{-4}\underline{M}$. The reaction can be monitored spectrophotometrically by noting the formation of NADPH (λ_{max} = 340 mμ).[1-4] Alternatively a chromogenic indicator reaction could be coupled with the enzyme reaction, and the amount of magnesium calculated by the rate of decrease of the color of a blue dye:

$$\text{NADPH + Oxidized Dye} \xrightarrow{\text{Diaphorase}} \text{NADPH + Reduced Dye}$$

  (blue)   (colorless) (2)

Or the reaction could be followed fluorometrically, by coupling with the resazurin indicator reaction (equation 3) as described by Guilbault et al[5]:

$$\text{NADPH + Resazurin} \xrightarrow[\text{Methosulfate}]{\text{Phenazine}} \text{Resorufin + NADP}$$

  (Non-Fluorescent)   (Fluorescent)   (3)

The rate of production of the highly fluorescent resorufin is proportional to the magnesium concentration.

The exact role of magnesium in the activation of ICDH is not known. However, it has been proposed that magnesium is involved in the binding of isocitrate to the NAD-specific enzyme. The interaction occurs at two interacting substrate sites, NAD binding at another.[6,7]

A thorough study of this reaction was made by Adler, von Euler, Gunther and Plass[2] and by Blaedel and Hicks[3] who found that both Mn^{2+} and Mg^{2+} efficiently activate ICDH. The analytically useful range for Mn^{2+} ranges from 5-100 parts per billion (Fig. 1). A study of interferences by Blaedel and Hicks[3] and Kratochvil et al[4] revealed that many metals inhibit the activation of ICDH. At $10^{-6}M$ Ag^+, Ce^{+3}, and Hg^{2+} and at $10^{-4}\underline{M}$ Ba^{2+}, Al^{3+}, Sr^{2+}, Pb^{2+}, Cu^{2+}, Ca^{2+}, Be^{2+}, Fe^{2+}, Ni^{2+}, Cd^{2+}, Th^{4+} and UO_2^{2+} all show inhibition ranging from 15-95%. Concentrations of In^{3+}, Pb^{2+}, Ag^+ and Hg^{2+} at $10^{-5}\underline{M}$ completely inhibit the ICDH.

G

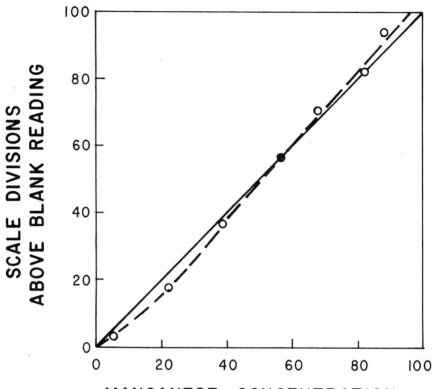

FIG. 1

Calibration curve for manganese determination
by ICDH activation.   Scale divisions are the
readings of absorbance of NADPH produced in
equation 1.   (Redrawn from Ref. 3).

——————————   =   plot expected.

- - - - - - - - - -   =   plot obtained.

Kratochvil, Boyer and Hicks[4] found that in addition
to manganese and magnesium, both zinc and cobalt are
activators of ICDH. Figure 2 indicates the relative
extent of activation, expressed as the rate of formation
of NADPH in the dehydrogenase reaction, (equation 1),
as a function of metal concentration for these four
metals. At 10^{-5}M zinc is a stronger activator than
Mg^{2+}, but higher concentrations of zinc effect a decrease
in the activation. At 3×10^{-4}M zinc becomes an inhibitor
of ICDH. Manganese(II) is the best activator of ICDH,
cobalt(II) the poorest of the 4 metals studied. Procedures for the analysis of trace amounts of these metals
were described.[4]   Analysis was performed either by a
measurement of reaction rates in the presence of a non-
limiting excess of reactants, or, for all but magnesium
by titration with EDTA.

A NAD specific ICDH has been isolated from Baker's
yeast which is activated by Co^{2+}, Mn^{2+} and Zn^{2+}.[8]
Hayakawa[9] has found that both Mg^{2+} and Mn^{2+}
activate heart pyruvic dehydrogenase and bacterial 2-
oxyglutarate dehydrogenase. Mg^{2+} and Mn^{2+} also have an
activating effect upon the DNA-DNase enzyme system and
can be determined.[10]

Magnesium can also be determined by its activation
of the firefly reaction:

$$\text{Luciferin} + O_2 + \text{ATP} \xrightarrow[Mg^{2+}]{\text{Luciferase}} \text{Oxyluciferin} + \text{ADP} + PO_4^{3-} \quad (4)$$
$$\text{(Green chemiluminescence)}$$

This reaction is discussed in section 3 below. As
little as 10 ppb of Mg^{2+} is determinable.[11]

2.   Determination of Barium

Townshend and Vaughan[12] have described a method for
the determination of 14-126 μg of barium in the presence
of calcium or magnesium. The method is based on the

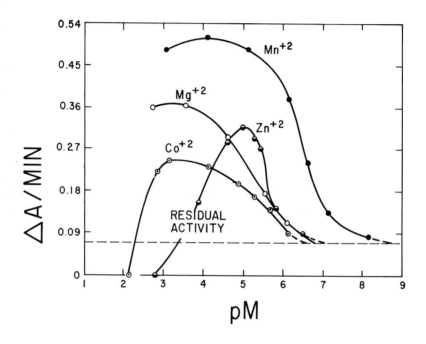

FIG. 2

Rate of formation of NADH vs. (-)
log metal concentration for manganese,
magnesium, zinc and cobalt (ref. 4).

reactivation of calf-intestinal alkaline phosphatase
that is inhibited by zinc. Some typical results
obtained for the assay of barium in the presence of
calcium and magnesium are indicated in Table 1.

TABLE 1

Determination of Barium

Barium, μg		Other Cation Added	
Present	Found	Cation	Conc, μg
14	14,17	--	--
14	12	Ca	42
42	42,42	--	--
70	72,70	--	--
70	73	Mg	25
98	96,95	--	--
98	97	Ca	42
126	126,120	--	--
126	120	Mg	25

Magnesium, calcium, strontium and manganese also
activate the enzyme and counteract the inhibition by
zinc. It is necessary, therefore, to attempt to mask
these ions without altering the effects of barium and
zinc, and without removing the zinc that is an integral
part of the enzyme. The authors found that a high
concentration of fluoride enabled barium to be determined in the presence of 50 μg of magnesium or calcium;
92 μg of strontium still re-activated the enzyme, but
18 μg was without effect. Many other metals which
either inhibit or re-activate the enzyme, including
lead, vanadium, aluminum, nickel and beryllium, interfere when present in μg amounts.

p-Nitrophenylphosphate must be used as substrate for
phosphatase for Ba^{2+} activation; no activation was

observed when α-naphthyl phosphate was used as substrate.

3.   Determination of Oxygen

   The mechanism of the firefly reaction has been elu-
cidated, and the luminescent reaction has been performed
in vitro by mixing cell-free extracts and even pure
reactants.   The firefly reaction requires ATP and Mg^{2+}
in addition to luciferin, luciferase and oxygen (equation 4).[11]   Ordinary fluorometric equipment can be
easily converted for use with chemiluminescent reactions
so that one can then measure the rate of formation of
luminescence (or the peak luminescence) and use this to
determine Mg^{2+}, O_2 or ATP.

   The firefly reaction may be used to assay oxygen at
partial pressures below 10^{-3}mm[13], when the gas is
passed through a bacterial emulsion containing all
requirements for the chemiluminescent reaction except
oxygen.

4.   Determination of Cyanide, Sulfide and Iodide

   Mealor and Townshend[14] have proposed analytical
methods for the assay of cyanide, sulfide and iodide
based upon the reactivation of an inhibited enzyme.

   The enzyme invertase was found to be inhibited in
its catalysis of the hydrolysis of substrates, such as
sucrose, by the presence of traces of certain metal ions
($2 \times 10^{-7}\underline{M}$ Ag^+ and $2 \times 10^{-8}\underline{M}$ Hg^{++}).   Equally small
concentrations of anions that form very strong complexes
with these two metals - cyanide, sulfide, iodide, etc. -
compete with the enzyme for the metals and decrease the
inhibitory effect.   Less than 1 ppm of iodide and 0.1 ppm
of cyanide or sulfide could be determined on the basis
of this de-inhibition or reactivation effect.

5.   Other Substances

   Table 2 summarizes the inorganic substances which have
been or could be determined by their activation of

enzyme systems.

TABLE 2

Activation of Enzyme Systems by Inorganic Substances

Substance	Enzyme System	Reference
Ba^{2+}	Alkaline phosphatase[a]	12
Ca^{2+}	2-Oxyglutarate dehydrogenase	9
	Taka-amylase	15
CN^-	Invertase[a]	14
Co^{2+}	Isocitrate dehydrogenase	4,8
I^-	Invertase[a]	14
K^+	Phosphofructokinase	16
Mg^{2+}	Isocitrate dehydrogenase	1-4,8
	Luciferase (Firefly)	11
	2-Oxyglutarate dehydrogenase	9
	Pyruvate dehydrogenase	9
	DNase	10
	Creatine phosphokinase	39,40
Mn^{2+}	Isocitrate dehydrogenase	1-4,8
	DNase	10
	Pyruvate dehydrogenase	9
	2-Oxyglutarate dehydrogenase	9
O_2	Luciferase (Firefly)	13
S^{2-}	Invertase[a]	14
Sr^{2+}	Taka-amylase	15
Zn^{2+}	Isocitrate dehydrogenase	4,8

[a]Methods based on the activation of an inhibited enzyme.

Takagi and Isemura[15] found that Ca^{2+} was needed for activation of taka-amylase A. A $10^{-4}\underline{M}$ concentration was needed for 80% regeneration, and only Sr^{2+} could replace Ca^{2+} in the activation. Dvornikova et al[16] found that the potassium salts of fructose-6-phosphate and ATP activate unpurified phosphofructokinase from rabbit muscle considerably more than Na^+, and proposed this activation as a method for K^+ in the presence of Na^+.

Hayakawa found that Ca^{2+} activates bacterial 2-oxy-glutarate dehydrogenase and thus can be determined.[9]

C. DETERMINATION OF COENZYMES

1. General

A number of enzymes require for their activity a specific coenzyme which participates in the enzymic reaction. By measuring the amount of activation of such an enzyme by the coenzyme a plot of initial rate of reaction vs. coenzyme concentration may be constructed. At low concentrations of coenzyme, the degree of activation will be proportional to the concentration of coenzyme added. Table 3 lists several of the coenzymes which are determinable together with their respective enzyme system.

2. Flavine Mononucleotide

Flavine mononucleotide (FMN) is a cofactor for the enzyme lactic oxidase from pneumococci:

$$\text{Lactate} + O_2 + FMN \longrightarrow \text{acetate} + CO_2 + H_2O + FMNH$$
$$(5)$$

The FMN content of a sample is determined by the activation of the enzyme.[17] The reaction is usually monitored by following the oxygen uptake in a Warburg manometer. A plot of rate of oxygen uptake vs. FMN concentration is linear up to about $10^{-7}\underline{M}$ FMN.

Alternatively FMN can be assayed using NADP and cytochrome C reductase.[18] The change in NADPH concentration,

TABLE 3

Enzyme Systems Used in Assay of Coenzymes

Coenzyme	Enzyme System	Reference
ADP	Pyruvate Kinase	34,37,38
AMP	Myokinase	34,37,38
ATP	Luciferase	27-32
	Hexokinase	33-35
	Phosphoglycerate Kinase	34,35
Creatine Phosphate	Creatine Phosphokinase, Hexokinase	39,40
	Creatine Phosphokinase, Luciferase	42,43
FAD	D-amino Acid Oxidase	19,20
FMN	Lactic Oxidase	17
	Cytochrome c Reductase	18
NAD	Alcohol Dehydrogenase	22,23
	Glutamate Dehydrogenase	26
NADH	Lactate Dehydrogenase	24
	Diaphorase	23
NADP	Glucose-6-Phosphate Dehydrogenase	24
	Glutamate Dehydrogenase	26
	Isocitric Dehydrogenase	25
NADPH	Glutamate Dehydrogenase	24
	Glutathione Reductase	25

measured spectrophotometrically at 340 mμ, is used to monitor the reaction.

3.   Flavine Adenine Dinucleotide

Flavine adenine dinucleotide (FAD) is the coenzyme of D-amino acid oxidase from pig kidney, and can be determined specifically by its activation of this enzyme:

$$\text{D-alanine} + \text{FAD} \xrightarrow{\text{D-amino acid}}_{\text{oxidase}} \text{pyruvic acid} + \text{FADH} +$$

$$\text{NH}_3 \qquad (6)$$

$$\text{FADH} + \text{O}_2 \longrightarrow \text{FAD} + \text{H}_2\text{O}_2$$

Warburg and Christian[19] and Straub[20] were the
first to describe this method for FAD. The oxygen
uptake, measured manometrically, is proportional to
the FAD in concentrations up to 25 μg/ml. The activation of D-amino acid oxidase by FAD must be compared
with standard solutions since the Michaelis constant
for the enzyme-FAD complex varies with temperature and
with the preparations of enzyme. Alternatively, the
reaction can be monitored fluorometrically using the
p-hydroxyphenylacetic indicator reaction described by
Guilbault and Hieserman[21]:

$$\text{H}_2\text{O}_2 + \text{p-hydroxyphenylacetic acid} \xrightarrow{\text{Peroxidase}}$$
$$\text{(non-fluorescent)} \qquad \text{Fluorescent Product} \quad (7)$$

The rate of production of fluorescence with time,
ΔF/min, is proportional to the FAD concentration.

4. Nicotinamide Adenine Dinucleotide (NAD), and Its
    Reduction Form (NADH)

NAD can be determined by quantitative reduction to
NADH by ethanol and alcohol dehydrogenase:

$$\text{Ethanol} + \text{NAD} \xrightarrow{\text{alcohol}}_{\text{dehydrogenase}} \text{NADH} + \text{acetaldehyde} + \text{H}^+$$
$$(8)$$

The equilibrium constant of this reaction favors the
oxidation of NADH to NAD, but the reduction of NAD
can be effected by using a pH of 9-10, a high ethanol
concentration, and by trapping the acetaldehyde formed
with hydrazine or semicarbazide.[22]   A pyrophosphate
buffer is generally used because pyrophosphate binds
heavy metal ions which may inhibit alcohol dehydrogenase.

The reaction can be monitored spectrophotometrically noting the increase in NADH at 340 mμ. Or the NADH can be measured fluorometrically at a λ_{ex} of 340 mμ, λ_{em} of 460 mμ. Similarly, the fluorometric resazurin indicator reaction of Guilbault and Kramer[23] can be used:

$$\text{NADH} + \text{Resazurin} \xrightarrow{\text{Phenazine methyl sulfate}} \text{Resorufin} \quad (9)$$
(non-fluorescent)   (fluorescent)

The rate of production of the highly fluorescent resorufin is proportional to the NAD concentration.

Reduced NAD, (NADH), can be assayed by any NAD-specific dehydrogenase reaction in which NADH is quantitatively oxidized. The reverse of the alcohol dehydrogenase reaction described above for assay of NAD can be used for NADH assay:

$$\text{Acetaldehyde} + \text{H}^+ + \text{NADH} \xrightarrow{\text{alcohol dehydrogenase}} \text{NAD} + \text{Ethanol} \quad (10)$$

Or the lactate dehydrogenase system can be used:

$$\text{NADH} + \text{H}^+ + \text{pyruvate} \xrightarrow{\text{lactate dehydrogenase}} \text{NAD} + \text{Lactate} \quad (11)$$

The equilibrium constants for both these reactions favor the oxidation of NADH even with only a small excess of substrate. The reaction can be monitored spectrophotometrically by noting the decrease in the absorbance of NADH.[24]

Muscle lactate dehydrogenase reacts with NADPH as well as with NADH, but this interference can be eliminated by working at a pH of about 7.8. At this pH the rate of oxidation of NADH is about 2000 times faster than with NADPH. Alcohol dehydrogenase preparations may contain small amounts of NADP-specific alcohol dehydrogenase, resulting in the oxidation of NADPH at high enzyme concentrations.

Guilbault and Kramer[23] described a fluorometric method for the analysis of $2 \times 10^{-7} \underline{M}$ to $2 \times 10^{-5} \underline{M}$ concentrations of NADH. The method is based on the production of the highly fluorescent resorufin (equation 9).

5.   Nicotinamide Adenine Dinucleotide Phosphate (NADP)
        and Its Reduced Form (NADPH)

NADP can be determined by its reduction to NADPH by glucose-6-phosphate dehydrogenase (G-6-PDH):

$$\text{Glucose-6-phosphate} + \text{NADP} \underset{\longleftarrow}{\xrightarrow{\text{G-6-PDH}}} \text{6-phosphogluconate} +$$
$$\text{NADPH} + \text{H}^+ \quad (12)$$

The equilibrium constant for this reaction favors the quantitative reduction of NADP, and the reaction can be monitored spectrophotometrically by the production of NADPH. The reaction is completely specific for NADP.[24]

Similarly, NADP can be assayed using the isocitric dehydrogenase system or any dehydrogenase system requiring NADP and not NAD (equation 1). The equilibrium for this reaction also favors the formation of NADPH, and the enzyme is specific for NADP.[25]

Reduced NADP, NADPH, can be assayed with the glutamate dehydrogenase system:

$$\alpha\text{-Oxoglutarate} + \text{NH}_4^+ + \text{NADPH} \xrightarrow[\text{Dehydrogenase}]{\text{Glutamate}}$$

$$\text{glutamate} + \text{NADP} \quad (13)$$

The equilibrium favors the formation of NADP, and the reaction can be monitored spectrophotometrically by the decrease in absorbance of NADPH.[24] Glutamate dehydrogenase is not specific for NADPH, but also reacts with NADH. The interference from NADH can be removed by oxidation with lactate dehydrogenase and pyruvate (equation 11).

NADPH can also be assayed with glutathione reductase, which is specific for NADPH:

Glutathione + NADPH + H$^+$ $\underset{\longleftarrow}{\overset{\text{Reductase}}{\longrightarrow}}$ 2-glutathione +
NADPH   (14)

The equilibrium is in favor of NADP formation, but
the activity of glutathione reductase is low compared
to glutamate dehydrogenase.[25]

6.   Enzyme Cycling Methods for NAD and NADP

Because of instrumental limitations, one is limited
to a sensitivity of about 10^{-8} moles per ml in spectrophotometry and 10^{-11} moles per ml with the more sensitive fluorometric methods in assay of NAD and NADP.

Lowry has proposed an enzymic cycling method for
measuring pyridine nucleotides[26], in attempt   to
increase this sensitivity limit by several orders of
magnitude.

The nucleotide to be assayed is made to catalyze
an enzymic reaction between two substrates, which are
transferred in amounts far greater than the nucleotide.
Thus, the measurement of the nucleotide through its
catalytic effect gives a 10^3 to 10^4 fold increase in
sensitivity over a direct measurement effect.   NAD is
measured with lactate dehydrogenase and glutamate
dehydrogenase:

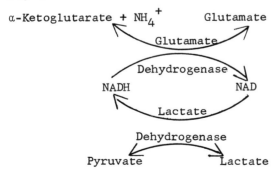

Pyruvate is produced in 2500 fold yield in 30 minutes and
is measured in a second cycle with added NADH and lactate
dehydrogenase.   The rate of transformation is measured
by following the change in either the fluorescent NAD or
NADH.   Since the nucleotides are used at concentrations
well below their Michaelis constants, the reaction rates
are proportional to the nucleotide concentrations.   The
final product is again a pyridine nucleotide, so the
cyclic process can be repeated with an overall multipli-
cation factor of 10^6 to 10^8.

The system for NADP measurement described by Lowry[26]
utilizes glucose-6-phosphate dehydrogenase and glutamate
dehydrogenase.   Each molecule of NADP catalyzes the
formation of up to 10,000 molecules of 6-phosphogluconate
in 30 minutes.   The 6-phosphogluconate is then measured
in a second incubation with 6-phosphogluconate dehydrogenase and extra NADP.   The NADPH produced is measured
fluorometrically.   Two cycle determinations have been
performed on as little as 10^{-15} moles of NADP.   This
detectable concentration represents the amount that would
be formed by 1000 molecules of an enzyme with a turnover

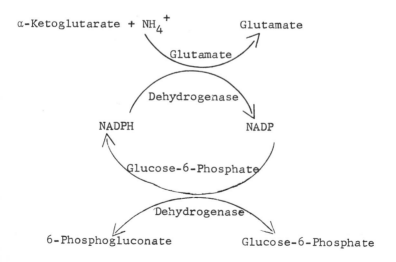

number of 10^4 per minute, if it could be coupled to an
NADP reaction. In principle, one could measure as
little as one single enzyme molecule by the reduction
of sample size or of the blank.[26]

7. Adenosine-5'-Triphosphate (ATP)

    a. <u>Assay with Luciferase</u>. The firefly reaction dis-
cussed above has been shown to require ATP in addition
to luciferin, luciferase, oxygen and Mg^{2+} (equation 4).
This reaction has been used as the basis for the most
sensitive method for ATP that is known.[27] Picogram
quantities of ATP were determined by Lyman and DeVincenzo[28] and Yokoyama and Nose.[29] Procedures for
ATP determination in red blood cells were proposed by
Beutler and Baluda.[30] In chemiluminescent analysis
a conventional fluorometer is used except that the
light source is blocked off from the sample cuvette so
that only light emitted from the firefly extract is
measured. Readings of peak luminescence are made after
about 1 minute and are proportional to the ATP concentration.

    This reaction is specific for ATP, although substances
which can alter the ATP concentration affect the luminescence. Aledort, Weed and Troup[31] found that the
light emission of the luciferin-luciferase reaction is
inhibited by increasing the ionic strength of the
medium. This decrease was proportional to the concentration of some cations such as Li^+, K^+ and Rb^+. A
comparison of the normal red cell ATP levels as measured by the firefly system and the hexokinase system
(see section b below) were made by Beutler and Mathai.[32]
Both methods were found to give the same results on
measuring the ATP content of trichloroacetic acid blood
filtrates, but protein was found to stimulate the light
output when lyophilized firefly extract was used. Thus,
when using the firefly method for ATP determination,
protein should be added to the standard.

b.   Assay with Hexokinase and Glucose-6-Phosphate
Dehydrogenase. Hexokinase catalyzes the ATP phosphorylation of glucose in the presence of Mg^{2+} to give
glucose-6-phosphate (equation 15). Glucose-6-phosphate
dehydrogenase (G-6-PDH) catalyzes the oxidation of
glucose-6-phosphate with NADP (equation 12).

$$\text{Glucose + ATP} \xrightleftharpoons{\text{Hexokinase}} \text{Glucose-6-phosphate + ADP} \quad (15)$$

The reaction can be monitored spectrophotometrically
(formation of NADPH).[33] Alternatively, the reaction
can be monitored fluorometrically by recording the
fluorescence of NADPH.[34,35] The increase in the
intensity of fluorescence is proportional to the ATP
concentration which can be calculated from a standard
calibration plot. Each mole of ATP forms 1 mole of
NADPH.

c.   Assay with Phosphoglycerate Kinase. Phosphoglycerate kinase catalyzes the reaction:

$$\text{3-Phosphoglycerate + ATP} \underset{Mg^{2+}}{\overset{\text{Kinase}}{\rightleftharpoons}} \text{1,3-diphosphoglycerate} + \text{ADP} \quad (16)$$

The 1,3-diphosphoglycerate is reduced by glyceralde-
hyde-3-phosphate dehydrogenase and NADH:

$$\text{1,3-Diphosphoglycerate + NADH + H}^+ \xrightarrow{\text{Dehydrogenase}}$$
$$\text{glyceraldehyde-3-phosphate} + \text{NAD + Phosphate} \quad (17)$$

Hydrazine is added to trap the glyceraldehyde-3-
phosphate and thus shift the equilibrium to the
right. The reaction is monitored by the decrease in
absorbance of NADH.[36] Alternatively, the decrease in
NADH can be measured fluorometrically.[34,35]

This procedure is specific for ATP. If myokinase

is present in the enzyme preparation, interference
from ADP will be observed because the following reaction occurs:

$$2\ ADP \xrightleftharpoons{Myokinase} ATP + AMP$$

8. Adenosine-5'-Diphosphate and Adenosine-5'-Mono-
    phosphate

Adenosine-5'-diphosphate (ADP) can be determined
by conversion to ATP with pyruvate kinase and 2-phos-
phonenolpyruvate:

$$ADP + phosphonenolpyruvate \underset{Mg^{2+},K^{+}}{\xrightleftharpoons{Kinase}} ATP + pyruvate \quad (18)$$

The pyruvate produced is then measured with lactate
dehydrogenase and NADH (equation 11).[34,37,38] The
rate of decrease in the absorbance of NADH, or the
change in fluorescence of NADH is proportional to
the amount of ADP present.

AMP does not interfere in this determination of
ADP, but creatine diphosphate, uridine diphosphate
and inosine diphosphate do and must be absent.

Adenosine-5'-monophosphate (AMP) can be determined
by phosphorylation with ATP in the presence of myokinase (MK). The ADP produced is determined as described above (equations 18 and 11).

$$AMP + ATP \underset{Mg^{2+}}{\xrightleftharpoons{MK}} 2\ ADP \quad (19)$$

Again, the rate of decrease in the absorbance at 366
mμ is a measure of the AMP present.

Thus mixtures of AMP and ADP, or ADP and ATP, can be
easily assayed.

9. Creatine Phosphate
    a. Hexokinase Procedure. Creatine phosphokinase
(CPK) catalyzes the transfer of phosphate from creatine phosphate to adenosine diphosphate (ADP).

$$\text{Creatine phosphate} + \text{ADP} \xrightleftharpoons[\text{Mg}^{2+}]{\text{CPK}} \text{creatine} + \text{ATP} \qquad (20)$$

The reaction can be monitored by coupling the hexokinase and glucose-6-phosphate dehydrogenase systems (equations 15 and 12).[39,40] The ATP produced phosphorylates glucose in the presence of hexokinase. The glucose-6-phosphate produced is oxidized catalytically by glucose-6-phosphate dehydrogenase in the presence of NADP. One mole of NADPH is liberated for each mole of creatine phosphate, and the increase in the absorbance due to NADPH produced is measured. The overall reaction is:

$$\text{Creatine Phosphate} + \text{Glucose} + \text{NADP} \xrightleftharpoons{\qquad}$$
$$\text{Creatine} + \text{6-phospho-}$$
$$\text{gluconate} + \text{NADPH} + \text{H}^+$$
$$(21)$$

The creatine phosphokinase procedure is specific for creatine phosphate. Inosine phosphates are practically inactive. In the reverse reaction ADP cannot replace ATP, and compounds related to creatine (creatinine or arginine) are not substrates.[41]

   b. Assay with Luciferase. The ATP formed from ADP in the creatine phosphokinase reaction (equation 20) can be assayed by the firefly reaction described in preceding sections (equation 4). The production of the chemiluminescence is a measure of the creatine phosphate present.[42,43] The peak luminescence produced after 30 seconds is linearly proportional to creatine phosphate in the 10-100 μg concentration region.

REFERENCES

1.   P. Baum and R. Czok, Biochem Z. _332_, 121 (1959).

2.   E. Adler, H. von Euler, G. Gunther and M. Plass, Biochem. J. _33_, 1028 (1939).

3. W. J. Blaedel and G. P. Hicks, "Advances in Analytical Chemistry and Instrumentation," (C. N. Reilley, ed.), Vol. 3, p. 118-120, Interscience, New York, 1964.

4. B. Kratochvil, S. L. Boyer and G. P. Hicks, Anal. Chem. 39, 45 (1967).

5. G. G. Guilbault, S. Sadar and R. McQueen, Anal. Chem., in press.

6. B. D. Sanival, C. S. Stachow and R. A. Cook, Biochemistry 4, 410 (1965).

7. M. Klingenberg, H. Goebell and G. Wenske, Biochem. Z. 341, 199 (1965).

8. C. Cennamo, G. Montecuccoli and G. Bonaretti, Biochim. Biophys. Acta 110, 195 (1965).

9. T. Hayakawa, Biochim. Biophys. Acta 128, 574 (1966).

10. R. Neske, Monatsber. Dent. Akad. Wiss Berlin 8, 675 (1966).

11. E. H. White, F. McCapra and G. F. Field, J. Am. Chem. Soc. 85, 337 (1963).

12. A. Townshend and A. Vaughan, Anal. Letters, 1, 913 1968.

13. A. M. Chase, in "Methods of Biochemical Analysis," Vol. 8, (D. Glick, ed.), Interscience, New York, 1960, p. 61.

14. D. Mealor and A. Townshend, Talanta 15, 1477 (1968).

15. T. Takagi and T. Isemura, Biochem. (Tokyo) 57, (1), 89 (1965).

16. P. D. Dvornikova, M. Gulyi and T. Pechenova, Ukr. Biokhim. Zh. 36, (6), 928 (1964).

17. S. Udaka, J. Koukol and B. Vennesland, J. Bacteriol. 78, 714 (1959).

18. E. Haas, B. L. Horecker and T. R. Hogness, J. Biol. Chem. 136, 747 (1940).

19. O. Warburg and W. Christian, Biochem. Z. 298, 150 (1938).

20. F. B. Straub, Biochem. J. 33, 787 (1939).

21. G. G. Guilbault and J. Hieserman, Anal. Biochem., in press.

22. E. Racker, J. Biol. Chem. 184, 313 (1950).

23. G. G. Guilbault and D. N. Kramer, Anal. Chem. 36, 2497 (1964).

24. J. Cooper, P. A. Srere, M. Tabachnik and E. Racker, Arch. Biochem. Biophysics 74, 306 (1958).

25. M. Klingenberg and W. Slenczka, Biochem. Z. 331, 486 (1959).

26. O. H. Lowry, J. V. Passonneau, D. Schulz and M. K. Rock, J. Biol. Chem. 236, 2746 (1961).

27. R. Wahl and L. Kozloff, J. Biol. Chem. 237, 1953 (1962).

28. G. E. Lyman and J. P. deVincenzo, Anal. Biochem. 21, 435 (1967).

29. S. Yokoyama and Y. Nose, Seikagaku 39, 46 (1967).

30. E. Beutler and M. Baluda, Blood 23, 688 (1964).

31. L. M. Aledort, R. I. Weed and S. B. Troup, Anal. Biochem. 17, 268 (1966).

32. E. Beutler and C. Mathai, Blood 30, 311 (1967).

33. A. Kornberg, J. Biol. Chem. 182, 779 (1950).

34. P. Greengard, Nature 178, 632 (1956).

35. P. Greengard, Photoelec. Spectrometry Group Bull. 11, 292 (1958).

36. H. J. Hohorst, F. H. Kreutz and T. Bucher, Biochem. Z. 332, 18 (1959).

37. H. Holmsen, I. Holmsen and A. Bernhardsen, Anal. Biochem. 17, 456 (1966).

38. F. Kubowitz and P. Ott, Biochem. Z. 314, 94 (1943).

39. A. C. Kibrick and A. T. Milhorat, Clin. Chim. Acta 14, 201 (1966).

40. L. Noda, S. A. Kuby and H. Lardy in "Methods in Enzymology," Vol. 2, (S. P. Colowick and N. O. Kaplan, eds.), Academic Press, New York, 1955, p.605.

41. M. L. Tanzer and C. Gilvarg, J. Biol. Chem. 234, 3201 (1959).

42. B. L. Strehler and J. R. Totter in "Methods of Biochemical Analysis," Vol. 1, Interscience, New York, 1954, p. 341.

43. B. L. Strehler and J. R. Totter, Arch. Biochem. Biophys. 22, 420 (1949).

CHAPTER 6

DETERMINATION OF INHIBITORS

A. GENERAL

An enzyme inhibitor is a compound that causes a decrease in the rate of an enzyme reaction, either by reacting with the enzyme to form an enzyme-inhibitor complex or by reacting with the enzyme-substrate inter- mediate to form a complex:

$$E + I \rightleftharpoons EI$$
$$E + S \rightleftharpoons ES \xrightarrow{I} E\text{-}S\text{-}I$$

In general, the initial rate of an enzymic reaction will decrease with increasing inhibitor concentration. This decrease will be linear at low inhibitor concen- trations, then will gradually approach zero. Several analytical methods have been proposed based on the in- hibition of an enzyme reaction and these will be dis- cussed in this chapter.

Analytical working curves for inhibitor assay are generally constructed by plotting % inhibition vs. con- centration of inhibitor. The % inhibition is calculated as follows:

$$\% \text{ Inhibition} = \frac{(\text{Rate})_{\text{No Inhibitor}} - (\text{Rate})_{\text{Inhibitor}}}{(\text{Rate})_{\text{No Inhibitor}}} \cdot 100$$

A control rate is recorded with no inhibitor present, but with the same volume of the solvent used to contain

the inhibitor added. This is especially critical in
studies of inhibitors added in non-aqueous solution,
since most non-aqueous solvents will inhibit the
enzyme in concentrations greater than about 3%. In
addition to the control (non-inhibited) rate, the
rate of spontaneous (non-enzymic) hydrolysis or oxidation of substrate should be measured and all rates
corrected for such non-enzymic effects, if necessary.

Generally a plot of % inhibition vs. concentration
is a typical exponential type curve[1], with a linear
region extending from 0 to 60 or 70% inhibition. This
linear region (Fig. 1) is the most analytically useful
range. If the inhibitor is a reversible rather than
irreversible one, then a curve such as that obtained in
Figure 1,B is obtained. The concentration ($\underline{M}$) of inhibitor that causes a 50% inhibition of the enzymic activity is the I_{50}, and is a measure of the strength of
an inhibitor.

Baker[2] and Bendetskii[3] have summarized the
theories of interaction of enzymes and inhibitors.
Lactic dehydrogenase, dihydrofolic reductase, thymidine
phosphorylase, guanase and xanthine oxidase are covered
in detail. Generally, the specific determination of one
metal ion inhibitor in the presence of others is imposs-
ible unless a prior separation or masking procedure is
used. The choice of masking agent will depend on a
number of factors: metals present, pH, the substrate
and enzyme, etc. Stehl, Margerum and Latterall[4]
have discussed the masking of metal ions in selective
rate methods. The use of masking agents (such as $S_2O_3^{2-}$
for Ag^+ or Hg_2^{++}; CN^- for Ni^{2+}, Co^{2+}, Fe^{3+}, etc.) will
allow the researcher to develop specific procedures for
very low concentrations of the metal ions of interest.

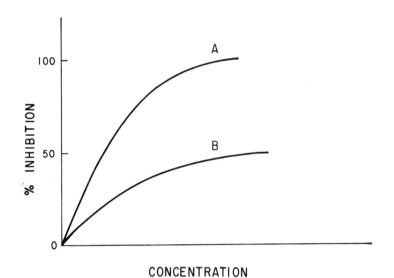

FIG. 1

Plots of % inhibition vs. concentration
for irreversible (A) and reversible (B)
inhibitors.

B.   DETERMINATION OF INORGANIC SUBSTANCES

1.   General

   Table 1 lists many of the inorganic substances that
have been determined based on their inhibition of an
enzymic reaction, together with the enzyme systems they
inhibit. The list is arranged with all the cations first,
anions second, and organic compounds last. Each group
is arranged alphabetically. Analytical methods for the
analysis of ions will be listed below by the enzyme systems inhibited in order to give the reader a feel for the
specificity of each system and the likely interferences
to be encountered.

TABLE 1

Substances Determinable Based on an Inhibition of an
Enzyme System

Substance	Enzyme System	Reference
Ag^+	Glucose Oxidase	5
	Invertase	8,11
	Isocitric Dehydrogenase	12
	Urease	19
	Xanthine Oxidase	20
Al^{3+}	Isocitric Dehydrogenase	12
Be^{2+}	Alkaline Phosphatase	17
Bi^{3+}	Alkaline Phosphatase	17
Ce^{3+}	Isocitric Dehydrogenase	12
Cd^{2+}	Isocitric Dehydrogenase	12
	Peroxidase	16
	Urease	19
Co^{2+}	Peroxidase	16
	Urease	19

TABLE 1 (Continued)

Substance	Enzyme System	Reference
Cu^{2+}	Hyaluronidase	7
	DNase	22
	Isocitric Dehydrogenase	12
	Peroxidase	16
	Urease	18,19
Fe^{2+}, Fe^{3+}	Hyaluronidase	7
	Isocitric Dehydrogenase	12
	Peroxidase	16
Hg^{2+}	Glucose Oxidase	5
	Glucosidase	6
	Invertase	8
	Isocitric Dehydrogenase	12
	Urease	19
	Xanthine Oxidase	20
In^{3+}	Isocitric Dehydrogenase	12
Mn^{2+}	Peroxidase	16
	Urease	19
Ni^{2+}	Isocitric Dehydrogenase	12
	Urease	18,19
Pb^{2+}	Glucose Oxidase	5
	Isocitric Dehydrogenase	12
	Peroxidase	16
	Urease	19
Zn^{2+}	Urease	18
CN^-	Hyaluronidase	7
$Cr_2O_7^=$	Peroxidase	16
F^-	Liver Esterase	13-15

TABLE 1 (Continued)

Substance	Enzyme System	Reference
$S^=$	Peroxidase	16
Hydroxylamine	Peroxidase	16
Ascorbic Acid	Catalase	66
p-chloromeruri-benzoic acid	Xanthine Oxidase	20
Cholesterol	β-Glucuronidase	68
DDT	Carbonic Anhydrase	50
Heparin	Pyruvate Kinase	59
	Ribonuclease	58,64,65
o-iodosoben-zoic acid	Xanthine Oxidase	20
Penicillin	D-Alanine Carboxypeptidase	70
Pesticides	Cholinesterase	23-49
	Lipase	51,53
	Phosphatase	17
Retinol	β-Glucuronidase	68
Thiourea	Invertase	11
Triton X-100	Lipase	51

2.   Glucose Oxidase

Toren and Burger[5] have studied the inhibition of glucose oxidase by three heavy metals, Ag^+, Hg^{++} and Pb^{++}. Methods for the microdetermination of these metals is based on the decreased rate of the enzyme catalyzed, aerobic oxidation of glucose to gluconic acid and hydrogen peroxide. In the presence of horse-radish peroxidase, the rapid oxidation of o-dianisidine

by hydrogen peroxide can be followed photometrically
at 440 mμ. Silver(I) was determined in the range of
5-200 ppb and mercury(II) in the range of 0.1-0.4 ppm.
Concentrations of lead(II) greater than 260 ppm are
needed for inhibition.

3.   Glucosidase

Guilbault and Kramer[6] found that glucosidase is
selectively inhibited by Hg^{2+} ions, and developed an
electrochemical method for the determination of this
ion in the 5×10^{-7} to $1 \times 10^{-5}\underline{M}$ concentration range.
The activity of the enzyme glucosidase was monitored
using the substrate amygdalin. Upon enzymic hydrolysis
cyanide is liberated which can be easily monitored using
silver and platinum electrodes in an internal (spontaneous) electrolysis cell. The rate of change of the
potential with time, $\Delta E/\Delta t$, due to the anode reaction:

$$Ag^{o} + 2\ CN^{-} \longrightarrow Ag\ (CN)_{2}^{-} + e^{-}$$

is proportional to the glucosidase concentration.

Mercury(II) inhibits the enzyme glucosidase, causing
a decrease in the amount of CN^{-} liberated and hence a
decrease in the slope of the potential-time curves
observed (Fig. 2). The following ions do not inter-
fere: SO_3^{2-}, NO_3^{-}, NO_2^{-}, F^{-}, Cl^{-}, Br^{-}, SO_4^{2-}, PO_4^{3-}
and Pb^{2+} (0.01M). Sulfide ion is an interference and
should be absent.

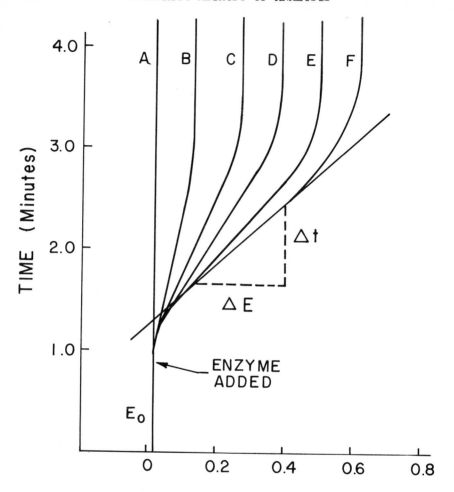

E, VOLTS

FIG. 2

Change in potential of a 0.0005 $\underline{M}$
amygdalin solution upon addition of
0.08 units of glucosidase (Almond
Emulsin).

A-E - Varying amounts of Hg^{2+} added
$\qquad$ $(5 \times 10^{-7}$ to $10^{-5}\underline{M})$.

F   - No Hg^{2+} added.

4.   Hyaluronidase

Guilbault and Kramer[7] have described a fluorometric method for the assay of the enzyme hyaluronidase and for Cu^{2+}, Fe^{2+} and CN^- ions which inhibit the enzymic activity.   The method for hyaluronidase is based upon the hydrolysis of the nonfluorescent indoxyl acetate by the enzyme to give the highly fluorescent indigo white (λ_{ex} = 395 mμ; λ_{em} = 470 mμ).

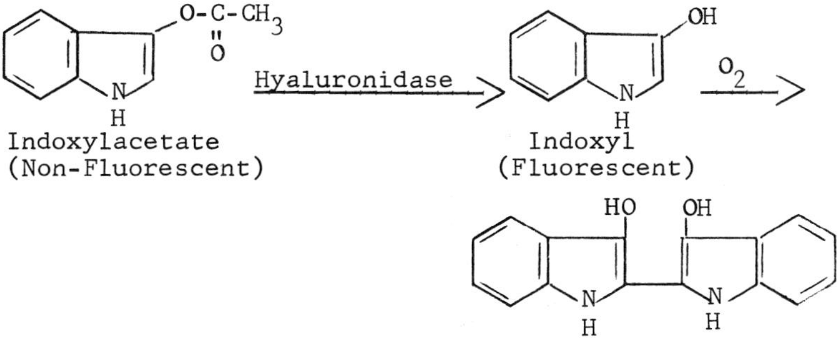

Indoxylacetate
(Non-Fluorescent)

Indoxyl
(Fluorescent)

Indigo White
(Highly Fluorescent)

The inorganic ions Fe^{2+}, Cu^{2+} and CN^-, which inhibit the enzyme, can be determined by recording their effect on the enzymatic activity.   From 0.10 to 4.0 μg/ml of CN^-, 0.20 to 12 μg/ml of Fe^{2+} and 0.20 to 6 μg/ml of Cu^{2+} can be determined with an accuracy and precision of about 2.3%.   Calibration plots of % inhibition vs. concentration were used to determine these 3 ions (Fig. 3).

The following ions had no effect on hyaluronidase and do not interfere: Pb^{2+}, Ag^+, Hg^{2+}, Al^{3+}, Cd^{2+}, Ni^{2+} and $S^=$.   No effect was noted from organophosphorus or chlorinated pesticides.

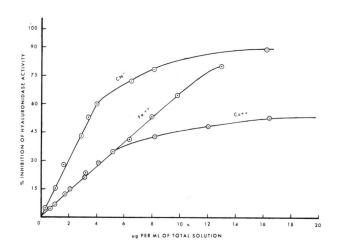

FIG. 3

Variation of percent inhibition
of hyaluronidase activity with
concentration of CN^-, Fe^{2+}, Cu^{2+}
(ref. 7).

5. Invertase

The enzyme invertase (β-fructofuranosidease) catalyzes the hydrolysis of sucrose to glucose and fructose. Mealor and Townshend[8] have described methods for the accurate determination of silver (2 to $10 \times 10^{-7}\underline{M}$) and mercury (2 to $10 \times 10^{-8}\underline{M}$) in the presence of each other and of most other metals based on the inhibition of the invertase catalysis of the hydrolysis of sucrose by these metals.

Mealor and Townshend[9] found that anions that bind strongly with these metals (cyanide, sulfide, iodide, etc.) compete with the enzyme for the metal ions, so that the inhibition is diminished. This effect was used in the determination of cyanide or sulfide (see Chapter 5, p.182). Numerous other complexing agents, such as histidine, methionine and reduced glutathione, have also been found to reduce the inhibitory effect of silver.[10] Mealor and Townshend[11] found that thiourea enhances the inhibition of invertase by silver ions and applied this effect to the determination of 1 to $5 \times 10^{-7}\underline{M}$ silver and of 10^{-7} to $10^{-8}\underline{M}$ thiourea.

6. Isocitric Dehydrogenase

Kratochvil, Boyer and Hicks[12] have reported the inhibition of isocitric dehydrogenase (ICDH) from pig heart by low concentrations of a number of cations: Ag^+, Al^{3+}, Ce^{3+}, Cd^{2+}, Cu^{2+}, Fe^{2+}, Fe^{3+}, Hg^{2+}, In^{3+}, Ni^{2+} and Pb^{2+} (Table 2). Analytical methods for these ions were either developed by the authors or proposed as possible. All rates were measured by following the NADPH formation at 340 mμ:

$$\text{Isocitric acid} + \text{NADP} \xrightarrow[\text{Mg}^{2+} + \text{Inhibitor}]{\text{ICDH}} \text{NADPH} +$$

$$\alpha\text{-keto glutarate}$$

TABLE 2

Effect of Various Ions on the ICDH System (ref. 12)

Ion	Concentration, $\underline{M}$	% Inhibition
Ag^+	3.3×10^{-7}	90
	3.3×10^{-8}	35
Al^{3+}	3.3×10^{-4}	100
	3.3×10^{-5}	43
	2.4×10^{-5}	22
Ce^{3+}	6.7×10^{-5}	80
	6.7×10^{-6}	65
	6.7×10^{-7}	22
Cd^{2+}	6.7×10^{-3}	100
	6.7×10^{-4}	56
	6.7×10^{-5}	5
Cu^{2+}	1.5×10^{-4}	81
	7.5×10^{-5}	41
	7.5×10^{-6}	5
Fe^{3+}	7.5×10^{-5}	50
	7.5×10^{-6}	12
Hg^{2+}	6.7×10^{-5}	100
	6.7×10^{-7}	57
	6.7×10^{-8}	3
In^{3+}	6.7×10^{-6}	100
	6.7×10^{-7}	69
	6.7×10^{-8}	23
Pb^{2+}	3.3×10^{-5}	100
	3.3×10^{-6}	61
	3.3×10^{-7}	13

The metal ions can be determined either by their inhibition of the rate of reaction or by a titration with
EDTA (EDTA complexes the inhibitor, thus effecting an
increase in the rate of reaction).

The authors[12] described procedures for the determination of mixtures of inhibiting metals, using EDTA
as a selective chelator.

Analysis of a mixture of Cu^{2+}, Mn^{2+} and Mg^{2+} can be
effected with an overall accuracy of 3% as follows: The
addition of EDTA causes an increase in the rate of reaction, ΔAbsorbance/min, since Cu^{2+}, a potent inhibitor,
has an EDTA formation constant greater than that of
manganese. A maximum rate is achieved after all the
Cu^{2+} has been complexed (Fig. 4). Then EDTA complexes
Mn^{2+} which is an activator of ICDH, so the rate falls.
The manganese is determined by extending the steeply
decreasing portion of the plot to the residual activity
level (which is the same as the dotted line obtained
with no Mg^{2+} added). Magnesium could not be determined
directly using EDTA as a titrant, but the peak observed
after the manganese end point gives a rate which could
be used in conjunction with a calibration plot to give
the Mg^{2+} concentration.[12]

7.   Liver Esterase

A highly selective method for the determination of
submicrogram amounts of fluoride was described by
Linde[13] and McGaughey and Stowell.[14] The method
is based upon the inhibition of the enzyme liver
esterase. The enzymic activity was measured by titration of butyric acid formed from ethyl butyrate with
NaOH.

Ethyl Butyrate $\xrightarrow[\text{Inhibitor}]{\text{Esterase}}$ Butyric Acid + Ethanol

Butyric Acid $\xrightarrow{\text{NaOH}}$ Sodium Butyrate + H_2O

H

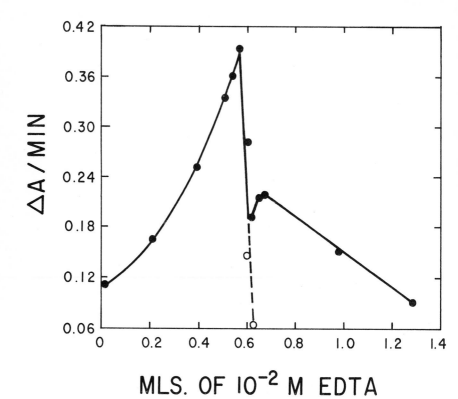

FIG. 4
Titration of Mn^{2+}, Cu^{2+} and Mg^{2+}
mixture with EDTA. Rate of forma-
tion of NADPH measured at 340 mμ
(ref. 12).

------- No Mg^{2+} added

The enzymic procedure overcomes the primary disadvantage of other methods in that no separation of fluoride from phosphate is required before analysis. Concentrations of phosphate greater than those normally found in body fluids ($10^{-3}\underline{M}$) were not found to interfere in the determination of nanogram concentrations of fluoride.[14] Maximum inhibition was observed in the 3-4 pH region, indicating that the probable inhibitor is not F^- but HF.[14]

McGaughey and Stowell also applied the liver esterase method to the detection of nanogram quantities of fluoride in tooth enamel.[15] No interference from ZrO_2 or Sn^{2+} was observed.

8. Peroxidase

Guilbault, Brignac and Zimmer[16] have described a fluorometric assay of the oxidative enzyme peroxidase based on the conversion of the nonfluorescent homovanillic acid(I) to the highly fluorescent 2,2'-dihydroxy-3,3' dimethoxybiphenyl-5,5'diacetic acid (II). The initial rate of formation of this fluorescent compound is measured and related to the activity of the enzyme:

I   II
(Non-Fluorescent)   (Fluorescent)

The inhibition of horse radish peroxidase by various inorganic substances necessary to give a 50% inhibition of the activity of peroxidase is noted in Table 3. Mn^{2+}, S^{2-}, Co^{2+}, $Cr_2O_2^{2-}$, Pb^{2+}, Fe^{2+}, Fe^{3+}, Cu^{2+}, Cd^{2+}, NH_2OH and CN^- are all good inhibitors of

TABLE 3

Concentration of Substances Causing a 50%
Inhibition of Horse Radish Peroxidase
(ref. 16)

Ion	Concentration, $\underline{M}$
Mn^{2+}	5.18×10^{-5}
S^{2-}	5.25×10^{-5}
Co^{2+}	7.59×10^{-5}
$Cr_2O_7^{2-}$	1.58×10^{-4}
NH_2OH	2.57×10^{-4}
Pb^{2+}	2.57×10^{-4}
Fe^{2+}, Fe^{3+}	3.16×10^{-4}
Cu^{2+}	6.46×10^{-4}
CN^-	6.76×10^{-4}
Cd^{2+}	4.17×10^{-3}

peroxidase and can be determined with good accuracy
($\sim$ 2%) and good precision ($\sim$ 3%)(Table 4). Results
obtained and reported with these substances are based
on the form of the substance added (and likewise to be
measured). Plots of percent inhibition vs. the log of
concentration of inhibitor were found to be linear over
the range 0-90% inhibition and over the following range
of inhibitors: 0.3-12.5 μg/ml of Mn^{2+}, 0.2-3 μg/ml of
S^{2-}, 0.5-25 μg/ml of Co^{2+}, 10-100 μg/ml of $Cr_2O_7^{2-}$,
10-185 μg/ml of Pb^{2+}, 5-65 μg/ml of Fe^{2+} or Fe^{3+},
5-120 μg/ml of Cu^{2+}, 0.9-125 μg/ml of CN^-, 50-200
μg/ml of Cd^{2+} and 3-85 μg/ml of NH_2OH.

TABLE 4

Determination of Inhibitors (ref. 16)

Cyanide, μg/ml		Sulfide, μg/ml[b]		Mn(II), μg/ml[c]	
Added	Found[a]	Added	Found[a]	Added	Found[a]
0.950	0.940	0.285	0.294	0.630	0.638
2.50	2.54	0.720	0.710	1.25	1.28
6.00	6.06	1.12	1.14	2.56	2.50
12.1	12.3	1.65	1.68	6.30	6.38
121.	118.	2.98	3.06	12.5	12.8
Av. Rel. Error	± 1.6	Av. Rel. Error	± 2.0	Av. Rel. Error	± 1.9

Hydroxylamine μg/ml[d]		Iron(II or III) μg/ml		Cobalt(II), μg/ml	
Added	Found[a]	Added	Found[a]	Added	Found[a]
3.10	3.20	5.18	5.15	0.530	0.541
8.01	8.17	12.8	13.0	1.27	1.29
14.6	14.8	25.4	26.0	2.60	2.55
30.1	29.8	49.5	49.0	5.30	5.32
80.0	81.8	61.3	60.0	12.7	12.4
Av. Rel. Error	± 2.1	Av. Rel Error	± 1.8	Av. Rel. Error	± 1.8

[a]Represents three or more determinations with a standard deviation of ± 3%.

[b]Mixture with 50 μg/ml of ClO_4^-, Cl^-, and NO_3^-.

[c]Mixture with 50 μg/ml of Mg^{2+} and Zn^{2+}

[d]Mixture with 50 μg/ml of Fe^{2+}, Ag^+, and Zn^{2+}. Results after ion exchange separation.

The following ions do not inhibit the enzyme and
were not found to interfere in analytical methods for
the inhibitors mentioned above: Na^+, K^+, Cu^{2+}, Mg^{2+},
Zn^{2+}, Hg^{2+}, Al^{3+}, Ag^+, ClO_4^-, F,Cl^-, Br^-, I^-, SO_4^{2-},
NO_3^-, PO_4^{3-} and nitride (N_3^-).

For samples containing a single inhibitory material,
with interfering materials either absent or previously
removed, a single rate measurement can be made, and the
concentration of inhibitor determined from a calibration
plot. Specificity can be built into the enzymatic determination of mixtures of inhibitors by the selective
use of chelating agents such as EDTA, citrate, F^-, etc.
Ion exchange techniques can likewise be used to separate
cationic and anionic inhibitors from each other and from
the nonionic inhibitors. Some of the results reported
in Table 4 were obtained from mixtures of various substances. For example, Mn^{2+} was assayed in the presence
of Mg^{2+} and Zn^{2+}; sulfide in mixtures of ClO_4^-, Cl^- and
NO_3^-; hydroxylamine in the presence of ions such as Fe^{3+},
and Zn^{2+} (which were removed before analysis by a cation
exchange resin)(Table 4).

9.   Alkaline Phosphatase

A fluorometric method was described by Guilbault,
Sadar and Zimmer[17] for the determination of the inorganic ions bismuth and beryllium, based on the inhibition
of akaline phosphatase by these ions. The substrate
umbelliferone phosphate was used, which is cleaved by
phosphatase to the highly fluorescent umbelliferone
(λ_{ex} = 365 mμ; λ_{em} = 450 mμ). Beryllium (0.01-0.30 μg/ml)
and bismuth (1-70 μg/ml) inhibit the hydrolysis cata-
lyzed by alkaline phosphatase, causing a decrease in the
slopes of the fluorescence-time curves, ΔF/min. This
decrease is a direct measure of the concentration of
inhibitor.

Table 5 lists a comparison of the I_{50} for various

TABLE 5

Comparison of I_{50} for Various
Inhibitors of Alkaline Phosphatase (ref. 17)

Inhibitor	I_{50}, $\underline{M}$	Incubation Time, min
Be^{2+}	1.65×10^{-5}	0
Bi^{3+}	1.5×10^{-4}	0

inhibitors of alkaline phosphatase. Of all the inorganic
ions tested, only Be^{2+} and Bi^{3+} had a strong inhibitory
effect on the enzyme. The anions sulfate, chloride,
bromide, iodide and phosphate do not interfere. Neither
do the cations Na^+, K^+, Li^+, Cu^{2+}, Mg^{2+}, Ni^{2+} and Mn^{2+}.
The following ions may be tolerated up to the concentra-
tions listed: fluoride (150 μg), dichromate (15 μg),
aluminum (15 μg), lead (150 μg), zinc (30 μg), copper
(30 μg), silver (10 μg), mercury (20 μg) and cadmium
(1 mg). No interference is observed in the determination
of beryllium or bismuth with these concentrations of
diverse ions present. Larger concentrations will cause
some decrease in the rate, dichromate and zinc being the
most serious interferences. [17]

10. Urease

Shaw[18] has reported that Cu^{2+}, Zn^{2+} and Ni^{2+} in
10^{-7} to $10^{-6}\underline{M}$ concentrations, and Co^{2+}, Fe^{2+} and Mn^{2+}
in 10^{-5} to $10^{-4}\underline{M}$ concentrations inhibit urease.

Toren and Burger[19] have described a kinetic, pH
stat method for the trace determination of eight metal
ion inhibitors based on the decrease rate of the urease-
catalyzed hydrolysis of urea. The order of inhibition
was $Ag^+ \rangle Hg^{++} \rangle Cu^{++} \rangle Cd^{++} \rangle Mn^{++} = Co^{++} \rangle Pb^{++} \rangle Ni^{++}$.

Silver(I) was determined in the range of 2 to 10 x 10^{-8} $\underline{M}$, mercury(II) in the range 2 to 10 x 10^{-7} $\underline{M}$, Cu^{++}, Cd^{++}, Co^{++} and Ni^{++} in the range of 2 to 10 x 10^{-6} $\underline{M}$.

11. Xanthine Oxidase

Guilbault, Kramer and Cannon[20] described an electro-chemical method for the determination of Ag^+ and Hg^{2+} based on the inhibition of the enzyme xanthine oxidase.

The method is based on the small current potentio-metric procedure discussed previously in Chapter 3 for xanthine oxidase and Chapter 4 for hypoxanthine and xanthine. The enzymatic reaction liberates peroxide which is electrochemically active. A depolarization of

$$\text{Hypoxanthine} + 2 \text{ H}_2\text{O} + \text{O}_2 \xrightarrow[\text{Inhibitor}]{\text{Xanthine Oxidase}} 2 \text{ H}_2\text{O}_2 +$$

Uric Acid

the platinum electrode results; the initial slopes of these depolarization curves $\Delta E/\Delta t$, are proportional to the xanthine oxidase concentration, as modified by the presence of Ag^+ or Hg^{2+}. From 0.54 to 3.2 μg/ml of Ag^+ and 1.0 to 10.0 μg/ml of Hg^{2+} were assayed with a relative error of $\pm$ 2%. Other cations that bind to sulfhydryl groups of the enzyme interfere in the deter-mination of Ag^+ and Hg^{2+} and must be absent (Cu^{2+}, Fe^{2+}, Zn^{2+}). No interference was observed from Bi^{3+}, Ni^{2+}, Mn^{2+} and PO_4^{3-}. Cyanide ion, which is electrochemically active and a known inhibitor of xanthine oxidase, is an interference.

The activity of xanthine oxidase can be monitored fluorometrically, in a method similar to that described above for peroxidase.[16]

12. Other Enzyme Systems

Igaue[21] has described the inhibition of the Q enzyme of rice plant by heavy metals such as Hg^{2+}, Ag^+, and

Cu^{2+} at 10^{-6} to $10^{-5}\underline{M}$ concentrations.

Neske[22] has reported that cattle pancrease DNase is inhibited by Cu^{2+}, thus providing a selective method for the determination of this ion.

C. DETERMINATION OF PESTICIDES

1. Anticholinesterase Compounds (Organophosphorus and Carbamate Compounds)

The most specific and sensitive method for some organophosphorus compounds is based on the inhibition of cholinesterase (ChE), and numerous papers have been published on this system. This inhibition of cholinesterase accounts to a large extent for the effectiveness of organophosphorus insecticides in the control of harmful insects in agriculture and for their toxicity to warm blooded animals. Kitz[23] has published a review with 159 references on the chemistry of anticholinesterase compounds, and Douste-Blazy[24] has compiled a comprehensive review of the enzymic hydrolysis of organophosphorus compounds.

The activity of cholinesterase, as modified by the pesticide present, can be assayed by any of the methods described in Chapter 3. In the method of Michel [25] the pesticide is extracted with organic solvent, the solvent evaporated off and the residue incubated with enzyme for 30 minutes with cholinesterase. At the end of the incubation time, substrate (acetylcholine) is added to the reaction mixture and the acetic acid produced is measured by the use of a pH stat approach:

Cholinesterase + Inhibitor $\longrightarrow$ Inhibited ChE + free ChE

Acetylcholine $\xrightarrow{\text{Free ChE}}$ acetic acid + choline

Acetic Acid + Base $\longrightarrow$ Sodium Acetate + HOH (constant pH maintained)

Alternatively, the free cholinesterase can be assayed electrochemically by the method of Guilbault, Kramer and Cannon.[26] The rate of the cholinesterase catalyzed hydrolysis of butyrylthiocholine iodide, as measured by dual polarized platinum indicator electrodes, is linearly related to the organophosphorus compounds (Fig. 5).

Anticholinesterase compounds inhibit the hydrolysis, causing a corresponding decrease in the slopes of the depolarization curves, $\Delta E/\Delta t$. This decrease is a direct measure of the concentration of organophosphorus compound. With 3-10 minutes preincubation as little as 2×10^{-4} $\mu g/ml$ of Sarin, 10^{-2} $\mu g/ml$ of Systox, 0.18 $\mu g/ml$ of parathion and 1.8 $\mu g/ml$ of malathion are determinable (Fig. 6).

Guilbault and Kramer[27] have shown that the use of a sensitive method like fluorescence allows the determination of much lower concentrations of enzymes. Since lower enzymic activities can be measured and lower concentrations of competitively inhibiting substrates used, lower concentrations of inhibitors might be determined.

Analytical methods for the determination of organophosphorus inhibitors based on their in vitro inhibition of cholinesterase were first described by Giang and Hall.[28] This method has been used successfully for compounds like Sarin, TEPP, Systox (P-O isomer) and paraoxan which are good in vitro inhibitors, but has been poor for the thiono- and dithio-phosphates which are weak in vitro cholinesterase inhibitors. This is unfortunate, since many of the important organophosphorus compounds are thiono- or dithio-phosphates.

Fallscheer and Cook[29] studied the conversion of some thiono- and dithio-phosphorus compounds to in vitro cholinesterase inhibitors. Dilute bromine water was found to effect the desired conversion to ChE inhibitors and excess bromine did not interfere with the enzymatic

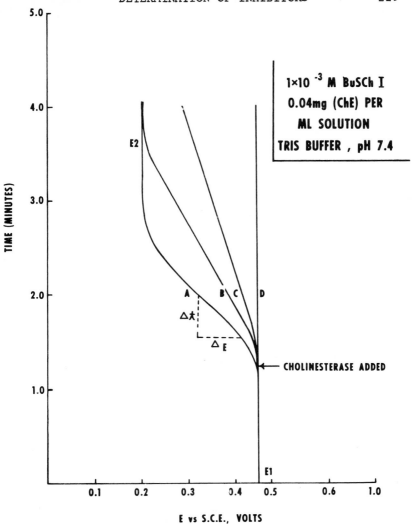

FIG. 5

Voltage-time curve for the enzymic
hydrolysis of BuSChI by ChE, inhibited
by Sarin (ref. 26).

    A.   No Sarin
    B.   0.36 μg/ml
    C.   0.50 μg/ml
    D.   1.0 μg/ml

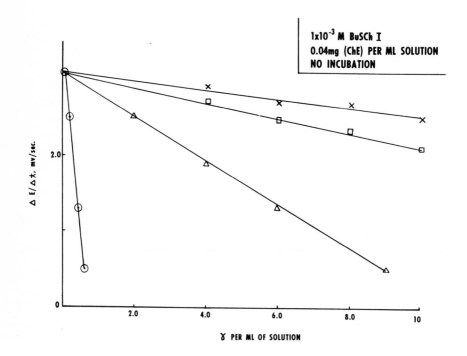

FIG. 6

Calibration plots of $\Delta E/\Delta t$ vs. inhibitor concentration ($\mu g/ml$) (ref. 26).

⊙   Sarin
△   Systox
□   Parathion
X   Malathion

reaction:

$$-\overset{|}{\underset{\downarrow}{P}}- \quad \xrightarrow{\text{Br}_2} \quad -\overset{|}{\underset{\downarrow}{P}}-$$
$$\quad\quad S \quad\quad\quad\quad\quad O$$

The reaction with most of the thionophosphates was
instantaneous after addition of bromine water and the
enzyme solution could be added immediately. Neither
Systox (P-S isomer) nor sulfatepp were converted to
inhibitors by bromine water but were slowly converted
by N-bromosuccinimide in $CHCl_3$ or CCl_4.

Other more difficultly oxidizable organophosphorus
compounds can be converted to inhibitors with H_2O_2-acetic
acid[30], nitric acid[28] or perbenzoic acid oxidation.[31]

Yurow, Rosenblatt and Epstein[32] found that many
monobasic acids of quinquevalent phosphorous form cholinesterase inhibitors when exposed to ketene.

Compounds of the type

$$\underset{R_1}{\overset{R}{>}}P\underset{X}{\overset{O}{<}}$$ where R and R_1 are alkyl or alkoxy

are generally active cholinesterase inhibitors if the
hydrolysis product HX has a pKa less than 7. Thus,
conversion of a monobasic acid of pentavalent phosphorus
to a derivative satisfying this requirement permits use
of an enzymatic method for assay of the organophosphorus
compound. Of 19 acids tested, 17 were easily converted
to cholinesterase inhibitors with ketene. Horse serum
cholinesterase was used for the detection studies, but
the authors reasoned that for the assay of special groups
of phosphorus acids it may be advisable to use different
enzyme sources.

In many cases a large increase in sensitivity can be

achieved by using another source of enzyme. In a study
of the inhibition of various enzymes by Sevin, Archer
and Zweig[33] found that some insect cholinesterases are
inhibited by much smaller concentrations of this carbamate than are other cholinesterases (Table 6).

TABLE 6

Comparison of I_{50} for Sevin with Various
Enzymes. All Conditions Optimum for
Analysis

Enzyme	I_{50}, µg
Cholinesterase, horse serum	1.8
Cholinesterase, bovine erythrocyte	4.0
Cholinesterase, human plasma	5.0
Cholinesterase, fly head	0.04

Thus fly head cholinesterase could be used for the
sensitive determination of Sevin.

Giang and Hall[28] assayed TEPP, paraoxan and other
insecticides that inhibit cholinesterase in vitro and
Kramer and Gamson[34] have developed a colorimetric
procedure with compounds related to the indophenyl
acetates for the determination of 1-10 µg of various
organophosphorus compounds. Underhay[35] has described methods for eserine and DFP using human red
cell plasma cholinesterase. Weiss and Galstatter[36]
detected pesticides in water by the inhibition of brain
cholinesterase in fish. Bluegill sunfish were found to
be the most sensitive.

Matousek and Cermon[37] have reported a highly sensitive, simple method for detecting cholinesterase inhibitors. Papers impregnated with butyrylthiocholine
iodide and bromthymol blue as a pH indicator were used.

Inhibition was indicated by the inability of the cholinesterase used to effect the hydrolysis of substrate.
Archer et al[38] reported a non-specific cholinesterase
inhibition procedure for the determination of ethion in
olives. Peracetic acid was used to oxidize the ethion
because olefinic compounds present in olives interfered
with the usual bromine treatment.[29] Cholinesterase
inhibition was used by Blumen[39] to determine phosdrin
in fruits and vegetables, and Winteringham and Fowler[40]
developed a method for Sevin based on the inhibition of
acetylcholinesterase. Abou-Donia and Menzel[41] studied
the inhibition of fish brain cholinesterase by carbamates
and developed an automatic method for the assay of this
class of pesticides.

Frequently, the pesticide to be determined is present
in a mixture, and hence must be separated from other
pesticides or other interfering compounds before analysis.
In many cases a preliminary extraction and/or a thin
layer chromatographic separation can be used to isolate
the desired compound for enzymic analysis. Mendoza et
al[42] have described an enzymatic inhibition method
sufficiently sensitive and reproducible for detecting
ten organophosphorus pesticides and carbaryl resolved
by thin layer chromatography (TLC). Reproducible reso-
lution of nanogram amounts of these pesticides is
achieved with a 450 μ thick gel layer, steer liver homo-
genate as source of esterase, and indoxyl or substituted
indoxyl esters as substrates. Mendoza, Wales, McLeod
and McKinley[43] proposed a rapid screening method for
organophosphorus pesticides in plant extracts without
an elaborate clean-up procedure. Malathion, parathion,
diazinon, ethion, etc. are extracted with acetonitrile
and partitioned into hexane before analysis by the thin
layer chromatographic (TLC)-enzymic inhibition technique. Similarly, Voss[44] and Ott and Gunther[45] pro-

posed a preliminary extraction of organophosphorus
pesticides prior to TLC separation, and Ott and
Gunther[46] in a later publication used the spots scraped
off a TLC plate as an input sample for the Auto Analyzer.
A test for the detection of organophosphorus pesticides
on the TLC plates using cholinesterase and either 2-
azobenzene-1-naphthyl acetate or indoxyl acetate was
described by Ortloff and Franz.[47] Ackerman[48] used
silica gel TLC plates for the semiquantitative determination of organophosphorus and carbamate pesticides
and Beam and Hankinson[49] reported a procedure for the
estimation of organophosphorus compounds in milk based
on cholinesterase inhibition.

DDT, Determination by Inhibition of Carbonic
Anhydrase

Keller[50] has reported a fairly specific method for
the determination of microgram amounts of DDT, based on
the inhibition of carbonic anhydrase. DDT inhibits this
enzyme at concentrations at which other inhibitors are
inactive.

Carbonic anhydrase catalyzes the hydration of carbon
dioxide to carbonic acid. The equilibrium lies in
favor of carbon dioxide, and reaction (2) is faster
than reaction (1).

$$CO_2 + H_2O \underset{\longleftarrow}{\overset{(1)}{\longrightarrow}} H_2CO_3 \underset{\longleftarrow}{\overset{(2)}{\longrightarrow}} H^+ + HCO_3^-$$

The standard method for assay of carbonic anhydrase
involves measurement of the CO_2 liberated by dehydration
in a Warburg manometer. DDT inhibits the enzyme, thus
causing a decrease in the rate of CO_2 liberation. The
concentration of DDT can be determined from a calibration
plot of % inhibition vs. concentration. As little as
10 μg of DDT is determinable with good reproducibility.
Below 10 μg the effect of impurities in the enzyme preparation becomes a critical factor, and possible non-

reproducibility or unmeasurable inhibition may result
from unspecific adsorption of DDT.

3.   Assay of Chlorinated and Other Pesticides Based on
     Inhibition of Lipase

Although considerable work has been performed in the
use of enzymes for the assay and determination of pesticides, with the exception of a few papers, all of past
research has been with cholinesterase, and very little
with other enzymes.

The objective of a research study conducted in the
author's laboratory was to attempt to develop sensitive
procedures for the determination of chlorinated insecticides, carbamate insecticides and herbicides, based on
the use of enzyme systems. Fluorescence methods were
used, because whenever tried such techniques have been
shown to allow the determination of much lower concentrations of enzymes and of enzymic inhibitors. The
enzyme system screened for inhibition was lipase, one
that Guilbault and Kramer[51] have reported to be inhibited
by 0.033 μg/ml of Sarin and Systox. A highly sensitive
fluorometric method, recently described by Guilbault and
Sadar[52], was used to monitor the lipase activity. The
non-fluorescent substrate, 4-methyl umbelliferone heptanoate,was used, which is cleaved by as little as 10^{-5}
unit of lipase to 4-methyl umbelliferone:

4-Methyl Umbelliferone Heptanoate $\xrightarrow{\text{Lipase}}$

     (Non-Fluorescent)
                              4-Methyl Umbelliferone

                                   (Fluorescent)

The chlorinated insecticides aldrin and lindane, and
the carbamate insecticide,Sevin,were found to be potent
inhibitors of lipase (Table 7), even without
incubation.[53]

TABLE 7

Comparison of I_{50} for Various Pesticides
Both With and Without Preincubation.
Lipase = 0.006 unit/ml

| Inhibitor | I_{50}, M[*] | | Optimum Incubation Time, min. |
	No Incubation	Preincubation[a]	
Aldrin	1.64×10^{-4}	1.93×10^{-5}	30
Sevin	4.0×10^{-4}	1.7×10^{-5}	10
Lindane	1.2×10^{-4}	6.18×10^{-5}	30
2,4-D	1.0×10^{-3} [b]	-----	--
Heptachlor	1.46×10^{-3}	4.24×10^{-5}	30
DDT	-----[c]	7.8×10^{-4}	20

[*]Concentration of pesticide giving 50% inhibition of enzymic activity.

[a]At optimum preincubation time.

[b]Preincubation has little effect.

[c]Reaches a maximum of 41% inhibition at 500 μg/ml, then decreases. Never reaches 50% inhibition.

Heptachlor and DDT, two other chlorinated pesticides, and the herbicide 2,4-D (2,4-dichloro-phenoxy acetic acid) were less potent,and larger concentrations were required for 50% inhibition of the enzyme (I_{50}).

Preincubation of enzyme and inhibitor before addition of substrate had a pronounced effect on aldrin, Sevin, lindane and heptachlor, but little effect on 2,4-D. Smaller concentrations of DDT effect a 50% inhibition (276 μg/ml compared to a maximum of 41% with 500 μg/ml with no preincubation), but DDT still remains a poor inhibitor. Optimum preincubation times are reported in Table 7.

Normal plots of percent inhibition vs. concentration
of inhibitor obtained were typical exponential type
curves. Plots of percent inhibition vs. the log of
concentration of inhibitor were found to be linear over
the range of 0-90% (Fig. 7), and over the following range
of inhibitors: 0.1-20 μg/ml of Sevin and aldrin, 0.5-
100 μg/ml of heptachlor, 0.8-100 μg/ml of lindane and
10-1,000 μg/ml of DDT and 2,4-D. Average errors and
precisions of 2-3% were obtained in all analyses, which
is quite acceptable for the low concentrations of pesticides analyzed.

The results indicate that the method of lipase inhibition is the most sensitive enzymic method yet reported
for heptachlor, aldrin, lindane and 2,4-D. Lipase is
more sensitively inhibited by Sevin than are bovine
erythrocytes and human plasma cholinesterase,[54] although
some insect cholinesterases are inhibited at smaller
concentrations of Sevin. The proposed inhibition scheme
is likewise a good one for DDT. Although carbonic anhydrase is inhibited by lower concentrations of DDT under
some conditions,[50] the lipase procedure described is
an easier, more convenient one to carry out. Instead
of an elaborate gaseous CO_2 monitoring system, only a
recording fluorometer is needed.

In order to eliminate possible interference from
inorganic ions, the pesticide can be extracted from the
sample with methyl cellosolve prior to analysis. Mixtures of pesticides can be separated by thin layer chromatography and each inhibitor determined separately.
Thus, the specific analysis of mixtures of pesticides in
samples of urine, crop materials, animal tissue, milk,
etc., is possible.[55]

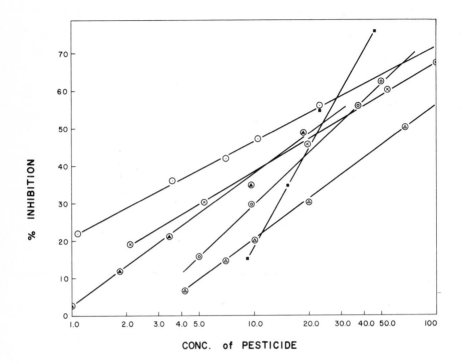

FIG. 7

Plot of percent inhibition vs. log of
concentration for a number of inhibitors
of lipase. Incubation times optimum.
Concentration scale varies with inhibitor.

⊙	Heptachlor,	1-100 μg/ml
Ⓧ	Sevin,	0.1-10 μg/ml
▲	Lindane,	1-100 μg/ml
⊚	DDT,	10-1,000 μg/ml
△	Aldrin,	0.1-10 μg/ml
■	2,4-D,	10-1,000 μg/ml

4.   Assay of Pesticides with Phosphatase
     Guilbault, Sadar and Zimmer[17] found that alkaline
phosphatase is inhibited by small concentrations of
aldrin and heptachlor in the 1-100 µg/range   The assay
procedure used is identical to that described above for
beryllium and bismuth. No other pesticides, either
organophosphorus, carbamate or chlorinated, were found
to inhibit phosphatase and do not interfere in the
determination.
     Parathion was found to specifically inhibit acid
phosphatase and can be determined in the 10-500 µg concentration range in the presence of other pesticides.

D.   DETERMINATION OF OTHER ORGANICS

1.   Sulfhydryl Binding Compounds
     Guilbault, Kramer and Cannon[20] have described an
electrochemical method for the assay of sulfhydryl-
attacking compounds, such as o-iodosobenzoic acid and
p-chloromercuribenzoic acid. These compounds inhibit
the enzyme xanthine oxidase causing a decrease in the
rate of depolarization of a platinum electrode in enzymic
action. Concentrations of 0.1 to 10 µg/ml can be determined. Metal ions that react with the active-S-H group
of the enzyme interfere in the determination. Alternatively, the fluorometric method of Guilbault et al[16]
can be used instead of the electrochemical procedure
for possible added sensitivity.

2.   Heparin
     Many enzyme systems are inhibited by heparin: acid
and alkaline phosphatase[56], hyaluronidase[57], ribonuclease[58], pyruvate kinase[59], gluatathione reductase[60], alcohol dehydrogenase[59], trypsin[61], fumarase[62] and adenyl deaminase[63] among others.
     Of all these, ribonuclease (RNase) and pyruvate
kinase are the most useful.

RNase cleaves ribonucleic acid, and cyclic 2',3'-pyrimidine nucleotides. The H^+ liberated is determined either by a pH stat type of approach or colorimetrically with a pH indicator. The rate of decrease in the H^+ liberated is proportional to the amount of inhibitor present. Alternatively, the rate of reaction can be monitored by noting the decrease in the absorbance of the solution at 300 mμ, due to degradation of the RNA by RNase, as inhibited by heparin.[64] Hobom and Zollner[65] detected as little as 0.5 μg of heparin by the inhibition of ribonuclease. Other mucopoly saccharides do not interfere.

The heparin inhibition of pyruvate kinase can similarly be monitored. In this case the spectrophotometric changes in NADH at 340 or 366 mμ are measured and heparin calculated from calibration plots of % inhibition vs. heparin concentration.

3. Ascorbic Acid

Catalase is inhibited by ascorbic acid in concentrations as low as $2 \times 10^{-6} \underline{M}$.[66] The rate of inhibition is increased by parts per billion concentrations of copper which does not inhibit alone. The activity of catalase can be monitored by any of the methods described in Chapter 3 for this enzyme (pp. 79-83).

4. Triton X-100

Guilbault and Kramer[51] have reported an analytical method for the determination of small concentrations of Triton X-100, based on the inhibition of the enzyme lipase.

A fluorometric method is used for the assay, and the amount of inhibitor determined from the decrease in the lipolytic hydrolysis of dibutyrylfluorescein to fluorescein.

5. Other Compounds

The inhibitory effect of D-cycloserine on catalase and peroxidase activity in plant and animal materials

was studied by Sicho and Kas[67], and the inhibition of
β-glucuronidase by cholesterol and retinol was assessed
by Tappel and Dillard.[68] A 70% inhibition of the
enzyme is effected by $10^{-4}\underline{M}$ retinol or $7.8 \times 10^{-5}\underline{M}$
cholesterol

Peroxylacetyl nitrate was determined in air pollution
studies by an inactivation of enzymes caused by an oxidation of enzyme sulflydryl groups and by inhibition of
the incorporation of acetate into fatty acids.[69]

Izaki and Strominger[70] reported that as little as
0.002 μg/ml of penicillin causes a 50% inhibition of the
enzyme D-alanine carboxypeptidase and can be determined.
Inhibition by other antibiotics (ampicillin, propicillin,
etc.) is also reported.

Mealor and Townshend[11] have described a method for
the determination of thiourea based on its enhancement
of the inhibition of invertase by silver ion. From 10^{-7}
to $10^{-8}\underline{M}$ thiourea was assayed, and a mechanism was
suggested for the enhancement.

REFERENCES

1. M. Dixon and E. C. Webb, "Enzymes," Academic Press,
    New York, 1958, pp. 171-180.

2. B. R. Baker, J. Chem. Ed. 44, 610 (1967).

3. K. M. Bendetskii, Dokl. Akad. Nauk. SSSR 171, 212
    (1966).

4. R. H. Stehl, D. Margerum and J. J. Latterell,
    Anal. Chem. 39, 1346 (1967).

5. E. C. Toren and F. Burger, Mikrochim. Acta 1968,
    538.

6. G. G. Guilbault and D. N. Kramer, Anal. Biochem.
    18, 313 (1967).

7. G. G. Guilbault and D. N. Kramer, Anal Biochem.
    18, 241 (1967).

8. D. Mealor and A. Townshend, Talanta 15, 747 (1968).

9. D. Mealor and A. Townshend, Talanta 15, 1477 (1968).

10. K. Myrback, Arkiv. Kemi. 11, 47 (1957).

11. D. Mealor and A. Townshend, Talanta 15, 1371 (1968).

12. B. Kratochvil, S. L. Boyer and G. P. Hicks, Anal. Chem. 39, 45 (1967).

13. H. W. Linde, Anal. Chem. 31, 2092 (1959).

14. C. McGaughey and E. Stowell, Anal. Chem. 36, 2344 (1964).

15. Idem., J. Dental Res. 45, 78 (1966).

16. G G. Guilbault, P. Brignac and M. Zimmer, Anal. Chem. 40, 190 (1968).

17. G. G. Guilbault, M. H. Sadar and M. Zimmer, Anal. Chim. Acta, Dec., 1968.

18. W. R. Shaw, J. Am. Chem. Soc. 83, 3184 (1961).

19. E. C. Toren and F. J. Burger, Mikrochim. Acta 1968, 1049.

20. G. G. Guilbault, D. N. Kramer and P. L. Cannon, Anal. Chem. 36, 606 (1964).

21. I. Igaue, Nippon Noeikagaku Kaishi 35, 1111 (1961).

22. R. Neske, Monatsber. Deut. Akad. Wiss. Berlin 8, 675 (1966).

23. R. J. Kitz, Acta Anaesthesiol. Scand. 8 (4), 197 (1964).

24. L. Douste-Blazy, Colloq. Nationaux Centre, Nat. Rech. Sci., Paris, 1965, 333.

25. H. O. Michel, J. Lab Clin. Med. 34, 1564 (1949).

26. G. G. Guilbault, D. N. Kramer and P. L. Cannon, Anal. Chem. 34, 1437 (1962).

27. G. G. Guilbault and D. N. Kramer, Anal. Chem. 37, 120 (1965).

28. P. A. Giang and S. A. Hall, Anal. Chem. 23, 1830 (1951).

29. H. O. Fallscheer and J. W. Cook, J. Assoc. Off. Agr. Chemists, 1956, 692.

30. G. G. Patchett and G. H. Batchelder, J. Agr. Food Chem. 8, 54 (1960).

31. P. A. Giang and M. S. Schechter, J. Agr. Food Chem. 8, 51 (1960).

32. H. Yurrow, D. Rosenblatt and J. Epstein, Talanta 5, 199 (1960).

33. T. E. Archer and G. Zweig, J. Agr Food Chem. 6, 910 (1958).

34. D. N. Kramer and R. M. Gamson, Anal. Chem. 29 (12), 21A (1957).

35. E. Underhay, Biochem. J <u>66</u>, 383 (1957).

36. C. M. Weiss and J. H. Galstatter, J. Water Pollution Control Fed. <u>36</u>, 240 (1964).

37. J. Matousek and J. Cermon, Procovni Lekorstvi <u>16</u>, 13 (1965).

38. T. E. Archer, W. L. Winterlin, G. Zweig and H. F. Beckman, J. Agr. Food Chem. <u>11</u>, 471 (1963).

39. N. Blumen, J. Assoc. Off. Agr. Chemists <u>47</u>, 272 (1964).

40. F. Winteringham and K. S. Fowler, Biochem. J. <u>99</u>, 6P (1966).

41. M. B. Abou-Donia and D. B. Menzel, Comp. Biochem. Physiol <u>21</u>, 99 (1967).

42. C. E. Mendoza, P. J. Wales, H. A. McLeod and W. McKinley, Analyst <u>93</u>, 34 (1968).

43. Ibid., p 173.

44. G. Voss, J. Econ. Entomol <u>59</u>, 1288 (1966).

45. D. E. Ott and F. A. Gunther, J. Assoc. Off. Agr. Chemists <u>49</u>, 662 (1966).

46. Ibid., p. 669.

47. R. Ortloff and P. Franz, Z. Chem. <u>5</u>, 388 (1965); Chem. Abs. <u>64</u>, 7304C (1966).

48. H. Ackerman, Nakrung <u>10</u>, 273 (1966); Chem. Abstr. <u>65</u>, 9657a (1966).

49. J. E. Beam and D. J. Hankinson, J. Dairy Sci , 1297 (1964).

50. H. Keller, Naturwissenschaften <u>39</u>, 109 (1952).

51. G. G. Guilbault and D. N. Kramer ,Anal. Chem. <u>36</u>, 409 (1964).

52. G. G. Guilbault and M. H. Sadar, Anal. Letters <u>1</u>, 460 (1968).

53. G G. Guilbault and M. H. Sadar, Anal. Chem., Feb., (1969).

54. T. E. Archer and G. Zweig, J. Agr. Food Chem. <u>6</u>, 910 (1958).

55. G. G. Guilbault and M. H. Sadar, in preparation for publication.

56. L. M. Buriuana, Naturwissenschaften <u>44</u>, 306 (1957).

57. M. B. Mathews and A. Dorfman, Physiol. Rev. <u>35</u>, 381 (1955).

58.   J. S. Roth, Arch. Biochem. Biophysics 44, 265 (1953).

59.   H. D. Horn and F. H. Bruns, Verh. Deutsch Ges. inn. Med. 65, 604 (1959).

60.   H. D. Horn and F. H Bruns, Biochem Z. 331, 58 (1958).

61.   M. K. Horwitt, Science 101, 376 (1945).

62.   A. Fischer and H. Herrmann, Enzymologia 3, 180 (1937).

63.   E. G. Dirnond, J. Lab. Clin. Med. 46, 807 (1955).

64.   N. Zollner and J. Fellig, Amer. J. Physiol. 173, 223 (1953).

65.   G. Hobom and N. Zollner, Z. Physiol. Chem. 335, 117 (1964).

66.   C. Orr, Biochim. Biophys. Res. Commun. 23, 854 (1966).

67.   V. Sicho and J. Kas, Sb. Vysoke Skoly Chem. Technol. V. Praze Potravin Technol 12, 5 (1966).

68.   A. L. Tappel and C. J. Dillard, J. Biol. Chem. 242, 2463 (1967).

69.   J. B. Mudd, Arch. Environ. Health 10, 201 (1965).

70.   K. Izaki and J. L. Strominger, J. Biol. Chem. 243, 3193 (1968).

CHAPTER 7

THE IMMOBILIZED ENZYME

A. GENERAL

One of the primary objections to the use of enzymes in chemical analysis is the high cost of these materials. A continuous or semicontinuous routine analysis using enzymes would require large amounts of these materials, quantities greater than can be reasonably supplied, and quantities that would require a prohibitive expenditure in many cases. If, however, the enzyme could be prepared in an immobilized (insolubilized) form without loss of activity, so that one sample could be used continuously for several hours or even days, and such that the enzyme could be stored at room or elevated temperature for months or years without any loss of activity, a considerable advantage would be realized. The immobilized enzyme could be used analytically in much the same way that the soluble enzyme is used, that is, to determine the concentration of a substrate, an inhibitor, or an activator.

Two major techniques can be used to immobilize an enzyme: (1) the chemical modification of the molecule by the introduction of insolubilizing groups. This technique resulting in a chemical "tieing down" of the enzyme, is in practice sometimes difficult to achieve because the insolubilizing groups can attach across the active site destroying the activity of the enzyme; (2) the physical entrapment of the enzyme in an inert matrix, such as starch or polyacryalmide gels. Physical entrapment techniques offer advantages of speed and ease of preparation. The major difference between the entrapped and the attached enzymes is that the former is isolated from large molecules which cannot diffuse into its matrix. The attached enzyme may be exposed to molecules of all sizes. Hence the two types of immobilized enzymes will differ in the form of the kinetics observed and in the kinds of interferences observed. Thus, for the assay of large substrates as proteins with proteolytic enzymes, an attached enzyme must be used and not an entrapped enzyme. Either enzyme could be used for the assay of small substrates such as urea.

B.   IMMOBILIZATION IN STARCH GEL

1. General

Vasta and Usdin[1] first showed that cholinesterase could be insolubilized by entrapment in a starch gel. Guilbault et al[2] found the procedure of Vasta and Usdin to be unsatisfactory because of a lack of satisfactory air and liquid flow and a lack of good air and liquid to enzyme contact. Preliminary experiments indicated that open cell polyurethane foam could be used as a support for the starch gel containing enzyme. Good reproducibility and uniformity were achieved in pad preparation, and the physically entrapped enzyme was shown to maintain its activity after 36 hours of operation.

2. Preparation of Immobilized Cholinesterase

Four grams of Connaught starch is placed into
10 ml. of 0.1M tris (hydroxymethyl) amino methane
buffer, pH 7.4, and the cool slurry is poured into
a boiling mixture of 28 ml of tris buffer and 2 ml
of U.S.P. glycerine. The resulting mixture is boiled
until a clear solution is obtained, after which it is
covered and allowed to cool to 47°C. In another
beaker, 400 mg of horse serum cholinesterase, (activity 3.0 units per mg; one unit = 1 μ mole of acetylcholine hydrolyzed per mg of enzyme per minute) is
dissolved in 5 ml of tris buffer, and this solution
is poured into the starch solution at 47°C (caution:
addition of enzyme at temperatures above 47°C will
cause denaturation of the enzyme). The beaker is
washed with 5 ml of tris buffer, making a total
volume of 50 ml. This enzyme starch solution is
gently stirred for 10 seconds, and immediately
poured onto 1 sq. ft. of 1/4 inch thick open cell
urethane foam (Scottfoam, Scott Paper Co., Phila-
delphia) which had been previously washed with
Alconox detergent, rinsed with distilled water, then
dried.

The enzyme-starch solution is gently worked into
the urethane foam with special care to minimize
foaming; then the urethane pad is gently squeezed
to remove the excess liquid. The pad is set in
the refrigerator at 40°F for an hour to gel, after
which it is placed in a vacuum desiccator (containing
no dessicant) and pumped overnight with a mechanical
pump. The large pad is cut into individual circular
pads of 3/4 inch diameter. Each pad contains approximately 12 mg of starch and 3 units of enzyme.

3. Properties and Use of the Immobilized Enzyme Gel
    The enzyme pads were cut to desired diameters with
a cork bore, with only gentle pressure being exerted.
The pads were assayed for enzyme content by soaking
the pad in water with squeezing to remove the enzyme.
Pads cut from the center had enzyme contents differing from those cut from the sides of the foam sheet
by only $\pm$ 5%. The pads were found to be stable at
room temperature for weeks, and could be used for
up to 36 hours for analysis. Using a flow rate of
0.5 ml/min, part of the enzyme (about 15%) is washed
off the pads in the first 15 to 20 ml of effluent,
and essentially no more is lost thereafter provided:
(1) mechanical stress is not applied to the pads
(squeezing, pressing, etc.) and (2) a low flow rate
of liquid is passed over the pads. Flooding of the
pads will cause eventual leeching of all enzyme from
the pads. Larger substrate flow rates will cause
more enzyme to be washed off the pads, as expected.
At a flow of 5 ml per minute, all of the enzyme is
washed off in about 4 hours; at 10 ml per minute,
the enzyme is lost in 1 hour. Up to 1 ml of liquid
per minute could be passed over the enzyme for
analysis without appreciable loss of activity.

    Addition of 5% glycerine to the starch gel pro-
duced pads that were less subject to mechanical
damage, and which were able to rehydrate more quickly
than pads without glycerine. Also, air or vacuum
drying of the impregnated pads produced pads that
were more uniform and higher in enzyme content than
those obtained from a freeze drying technique.

    The advantage of the insolubilized cholinesterase
lies in the fact that it, unlike the soluble enzyme,

is not used up in an analysis, but the same material
can be used for up to 36 hours of analysis.

To test the utility of the immobilized enzyme for
substrate analysis Guilbault et al[2] determined the
concentration of the substrates acetyl- and butyrl-
thiocholine iodide. A sample of the substrate was
passed over the enzyme, the effluent collected, and
the voltage recorded (pp. 33-34). From a calibration plot of potential vs. the logarithm of substrate
concentration, the amount of thioester present could
be calculated. Guilbault and Das[3] measured the
concentration of N-methyl indoxyl acetate[4] using
the immobilized cholinesterase pads. A sample of
the non-fluorescent ester is passed over the enzyme
pad, the effluent is collected and its total fluorescence measured. Typical calibration plots of
fluorescence vs. concentration of N-methyl indoxyl
acetate are shown in Fig. 1, A, B, C and D.

Guilbault et al have used the immobilized enzyme
pad to monitor continuously water and air for atmospheric pollutants which are enzymic inhibitors of
cholinesterase. Electrochemical[2] and fluorescence[5]
methods were used for the assay.

4. Automatic Monitoring of Inhibitors

a. Electrochemical Method. The electrochemical
apparatus used to monitor the activity of the enzyme
in the urethane pads continuously is indicated in
Fig. 2.[2] This figure shows the details of the
enzyme pad, O-ring and disc electrode assembly.
The disc electrodes were prepared by punching 1/16
inch diameter holes into a 1 inch circular piece of
0.003 inch thick platinum sheet that has a 3/8 x 1/4
inch handle (available on special order from J.

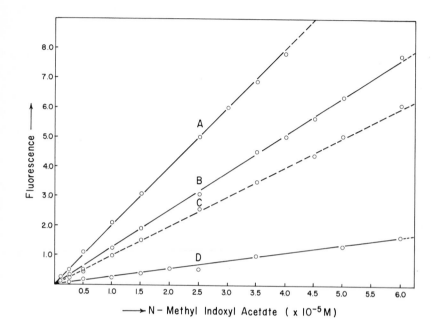

FIG. 1

Plot of total fluorescence obtained from
N-methyl indoxyl acetate hydrolyzed upon
passage over a number of gel immobilized
cholinesterase pads.

A - 9.7 units per pad
B - 6.5 units per pad
C - 4.4 units per pad
D - 0.52 units per pad

(One unit of enzyme catalyzes the hydrolysis
of one micromole of acetylcholine per mg
of enzyme per minute at 25°C)

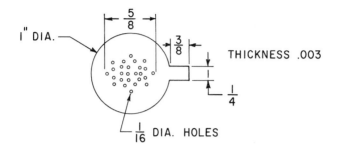

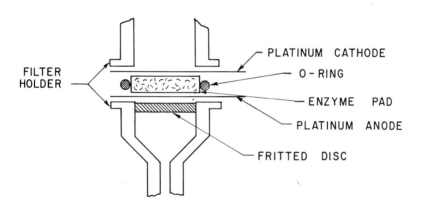

FIG. 2

Details of enzyme pad, O-ring, and grid
electrode assembly. (ref. 2)

Bishop and Co., Malvern, Pa.). A pad 3/4 inch in
diameter, prepared as described above, is then placed
into a 1 x 3/4 x 1/8 inch O-ring, the electrodes are
placed above and below the pad, and the pad and electrodes are placed into a Millipore micro analysis
filter holder.

The filter holder was held together with the
clamp provided with the filter, and the waste was
collected in a 250 ml filter flask. The substrate,
$5 \times 10^{-4}\underline{M}$ butyryl thiocholine iodide in $0.1\underline{M}$ tris
buffer, pH 7.40, is pumped over the pad using a
positive displacement liquid pump, with a delivery
rate of 1.0 ml/minute. Air and water (containing
possible cholinesterase inhibitors, i.e., pesticides)
were sucked through the enzyme pad by means of a
Brailsford blower. A constant current of 2 µa was
applied across the electrodes, and the change in
potential that occurs was monitored with a high
impedance electrometer, and was automatically re-
corded·

As long as the enzyme cholinesterase is active in
the pad, the butyrylthiocholine iodide will be hydrolyzed to the easily oxidizable thiocholine. At a
constant current of 2 µa, a potential of about 150 mV
(Fig. 3) will be established across the cell assembly
pictured in Fig. 2. Since the electrooxidation of the
thiol takes place at the anode, it is important that
the anode be located at the downstream surface of the
pad where the concentration of hydrolysis product is
greatest:

Acetyl Thiocholine Iodide $\xrightarrow{\text{ChE}}$ Thiocholine +

Acetic Acid

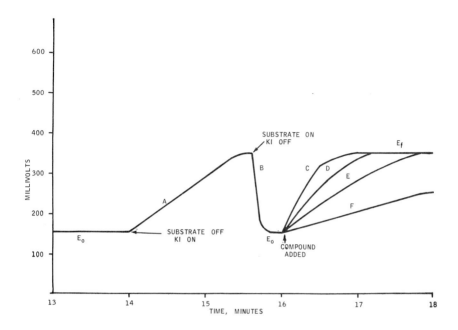

FIG. 3

Typical operation and response curves of experimental apparatus.

A. Flow rate 1 ml/min   Substrate solution, Off; KI solution, On.

B. Flow rate, 1 ml/min.   Substrate solution, On; KI solution, Off.

C. D, E and F.   Various concentrations of the pesticide systox added to water stream.

If an inhibitor is present in air or water that reduces
the activity of enzyme, less thiol will be formed, and
the potential will rise to that of the iodide/iodine
couple, 350 to 400 mV (Fig. 3, E_f). If the inhibitor
is a reversible one, it can be removed from the pad
by substrate flow, allowing subsequent determinations.
If the inhibitor is an irreversible one, a new
enzymic pad must be used.

   b. Fluorescence Method. A continuous fluorometric
system for the assay of anticholinesterase compounds
was designed and constructed by Guilbault and Kramer[5]
using the fluorescence attachment to the DK-2 spectrophotometer. A special glass tube was prepared (Fig. 4)
which fits into the fluorescence attachment to the
DK-2. This tube had a lower constriction, 10 mm wide
and 30 mm high, into which were stacked four enzyme
pads containing immobilized cholinesterase. The
wavelength of 320 mμ was employed for excitation,
using a Sylvania 360 Bl 4 W lamp (L) and a Corning
filter 72786 (F). At right angles to the pad was
placed the exit slit to the DK-2, and the emission
at 410 mμ was automatically recorded. Alternatively,
any fluorometer can be used with appropriate filters
to give a λ_{ex} of 320 mμ and a λ_{em} of 410 mμ. A
solution of 3 x 10^{-4}M 2-naphthyl acetate in Elving
buffer, pH 7.4, was passed over the pads at a rate
of 0.5 ml per minute using a Holter peristaltic
liquid pump; or air can be sampled over the enzyme
at a rate of 1 liter per minute using a Brailsford
blower at an air flow rate of 1 liter per minute
with up to 1 ml of liquid per minute. As long as
the cholinesterase is active, the 2-naphthyl acetate
is hydrolyzed to the highly fluorescent 2-naphthol,

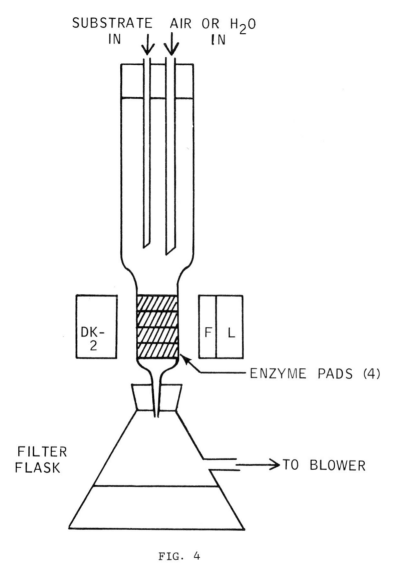

FIG. 4

Experimental fluorescence apparatus
(ref. 5)

and a high fluorescence is recorded (Fig. 5). When
the air or water becomes contaminated with an anticholinesterase compound, such as an insecticide, the
enzymic activity is blocked or lowered, less fluorescence is produced, and the fluorescence drops to
a low value. A drop in the base line fluorescence
indicates the presence of a contaminant, and the
rate of fall of the fluorescence with time provides
a semi-quantitative estimation of the concentration
of inhibitor, if the identity of this compound is
known. [5]

C.   IMMOBILIZATION IN POLYACRYLAMIDE GELS

1. General

An alternative to the starch gel method for
physical entrapment of enzymes lies in the use of a
polyacrylamide gel. Polyacrylamide gels were first
prepared by Bernfeld and Wan. [6] Hicks and Updike[7]
have trapped several enzymes in a polyacrylamide gel:
glucose oxidase, catalase, lactic dehydrogenase,
amino acid oxidase, glutamate dehydrogenase, and
enzyme activity in human serum. The preparation is
stable and can be conveniently stored. Details on
the preparation and properties of this gel are listed
below.

2. Preparation of Gels (Method of Hicks and Updike[7])

Forty grams of acrylamide (Eastman) is dissolved
in 100 ml of pH 7.4 phosphate buffer, 0.1 M. A
solution of the polymerizing agent is prepared by
dissolving 2.3 grams of N,N-methylene-bis acrylamide in 100 ml of phosphate buffer, 0.1 M, pH 7.4.
These solutions should be prepared fresh prior to
use. To prepare gels mix 1 ml of acrylamide solution

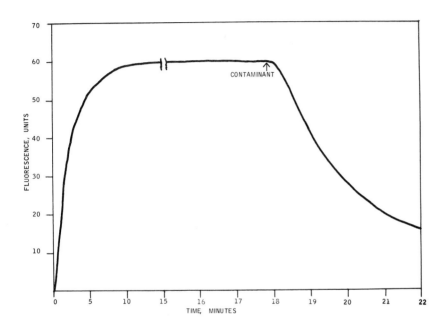

FIG. 5

Typical operation and response
curve for fluorescence apparatus.
(ref. 5)

with 4 ml of polymerizing agent solution and add 1 ml
of enzyme solution (containing about 10 mg of enzyme).
Start the polymerization reaction by adding 0.03 mg
of riboflavin and 0.03 mg of potassium persulfate
(which are the best catalysts for the polymerization
reaction). The system is deoxygenated by bubbling
with nitrogen before addition of enzyme. The copolymerization reaction is complete within 2-15
minutes using photocatalysis with a No. 2 Photoflood
lamp. The completeness of reaction is indicated by
maximum opacity. The copolymerization reaction proceeds nicely after oxygen is removed, and is usually
run in an ice bath to prevent the heat generated in
the reaction from denaturing the enzyme.

The resulting polymerized enzyme-gel was mechan-
ically dispersed into small particles, then lyphi-
lized, and sieved to a 20-40 mesh size.

The instrumentation Hicks and Updike[7] used to
study the immobilized enzyme is indicated in Fig. 6.
A column was packed with the enzyme-gel prepared as
described above, and substrate was passed over the
gel at a rate of 0.8 ml per minute. The column and
tubing preceeding the column were thermostated at
constant temperature. A portion of the column
effluent (S_1) (0.2 ml per minute) is mixed with a
stream of color reagent to detect the reaction products in the effluent stream. After passing through
a short delay line to permit the color reaction to
develop,

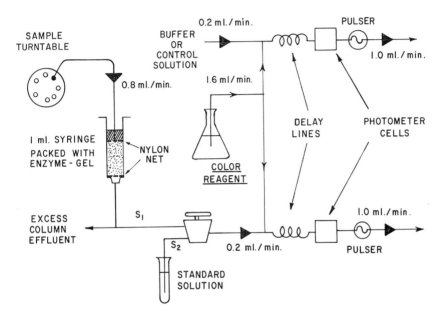

FIG. 6

Diagram of instrumentation used by
Hicks and Updike (ref. 7) for spectrophotometric studies of immobilized
enzymes.

a. Dehydrogenase System:

Substrate + NAD $\xrightarrow{\text{Dehydrogenase}}$ Product + NADH

NADH + Dye$_{(ox)}$ $\xrightarrow{\text{PMS}}$ NAD + Dye$_{(red)}$

(blue)   (colorless)

b. Oxidase Systems:

Substrate + O_2 $\xrightarrow{\text{Oxidase}}$ Product + H_2O_2

H_2O_2 + Dye$_{(red)}$ $\xrightarrow{\text{Peroxidase}}$ H_2O + Dye$_{(ox)}$

(colorless)   (blue)

the reaction stream passes to a photometer cell.
Simultaneously, the color reagent is mixed with buffer
or a standard control solution (S_2) and passed through
a second delay line into a photometer cell to serve
as a reagent blank. The difference in absorbance
between the two cells is a measure of the product
concentration in the column effluent stream. Standard solutions are introduced at S_2 to permit calibration.

3. Properties of the Gel

Hicks and Updike[7] have succeeded in immobilizing
enzymes in polyacrylamide gels prepared in the form
of particles, strings, blocks, tubes and coatings.

Best mechanical rigidity was obtained at high gel
concentrations over the range of concentrations
studied. At any one concentration of gel, an increase
in the amount of crosslinking agent decreases mechanical rigidity, but increases the yield of immobilized enzyme activity per unit of soluble enzyme
activity introduced. The most suitable gel material

requires both a relatively high concentration of
monomer to give mechanical rigidity and a high concentration of crosslinking agent to achieve the
highest possible yield of immobilized enzyme activity.

Hicks and Updike[7] prepared gels containing
various concentrations of enzyme to study the effect
of enzyme concentration on the enzyme gel activity.
Some typical results obtained with glucose oxidase
columns ranging from 5 to 420 mg of glucose oxidase
per 100 ml of gel are indicated in Fig. 7. Glucose
was passed over the immobilized enzyme column, and
the H_2O_2 produced was measured colorimetrically and
found to be proportional to the glucose concentration.
The slope of activity of gel (Δ H_2O_2 produced/Δ Glucose Oxidase) is proportional to the enzyme concentration up to 50 mg per 100 ml. Optimum activity
of the gel for analytical purposes, taking into
account the cost of enzyme, was found to be 25 mg
per 100 ml for glucose oxidase. Similar results
were obtained for other enzymes.[7]

The polyacrylamide enzyme gels were found to show
little loss of activity after 3 months of storage at
$0-4^{\circ}C$. Lactate dehydrogenase (LDH) lost about 30%
of its activity in 3 months while glucose oxidase
(GO) showed no loss in activity. Hydration of a
LDH gel caused a loss of all activity in 3 months
at $4^{\circ}C$, whereas a hydrated GO gel exhibited a loss
of only 5% at $0^{\circ}C$ in 3 months. The gels are very
resistant to flow loss, and were used for the assay
of the substrates glucose and lactic acid.[7]

To test the stability of the immobilized enzyme
as compared to the soluble enzyme, the activity of
a series of identical glucose oxidase columns and

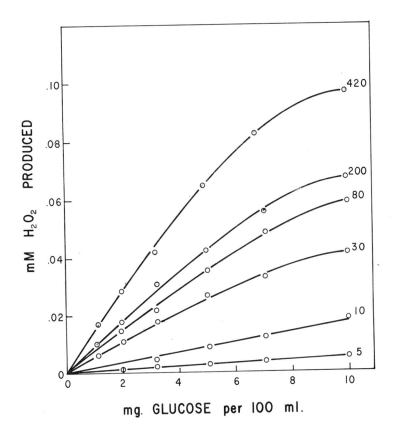

FIG. 7

Effect of enzyme concentration on
column responses. Various curves
represent total mg of Glucose Oxidase used per 100 ml of Gel (ref. 7).

glucose oxidase solutions were compared after heating for 10 minutes at temperatures of 37-70°C. About half of the activity of GO in both gel and solution was destroyed in 10 minutes at 60°C and all of the activity of both were lost at 70°C.

Updike and Hicks have coupled the immobilized glucose oxidase system with an electrochemical sensor for the determination of glucose in blood.[8] The oxygen electrode, described in Chapter 2, was used to monitor the oxygen uptake:

$$\text{Glucose} + O_2 \xrightarrow{\text{Glucose Oxidase}} H_2O_2 + \text{Gluconic Acid}$$

The apparatus used to monitor the reaction is indicated in Fig. 8. Immobilized glucose oxidase, prepared as described above, is placed in a miniature chromatographic column, and samples containing glucose to be analyzed are pumped over the column at a rate of 0.4 ml/min with a peristaltic pump. Using an immobilized enzyme and an oxygen electrode as the sensing device, a "reagentless" analyzer was achieved (Fig. 8).

In a further development of the immobilized enzyme concept, Updike and Hicks[9] described the preparation of an enzyme electrode. The electrode was a miniature chemical transducer which is prepared by polymerizing a gelatinous membrane over a polarographic oxygen electrode as shown in Fig. 9. When the enzyme electrode is placed in contact with a biological solution or tissue, glucose and oxygen diffuse into the gel layer of immobilized enzyme. The rate of diffusion of oxygen through the plastic membrane to the electrode is reduced in the presence of glucose and glucose oxidase by the enzyme

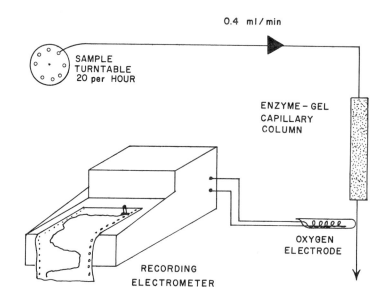

0.4 ml/min

SAMPLE
TURNTABLE
20 per HOUR

ENZYME – GEL
CAPILLARY
COLUMN

OXYGEN
ELECTRODE

RECORDING
ELECTROMETER

FIG. 8

Instrumentation system used by Updike
and Hicks (ref. 8) to monitor glucose
in blood.

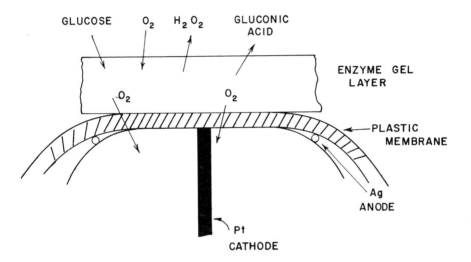

FIG. 9

Principle of enzyme electrode

(ref. 9)

catalyzed oxidation of glucose.

When the glucose concentration is well below the K_m for insolubilized glucose oxidase, and the oxygen is in non-rate-limiting excess there is a linear relationship between the reduction in oxygen content and the glucose concentration. Calibration curves of electrode response vs. glucose concentration are prepared (Fig. 10), and from these the amount of glucose present in whole blood or plasma can be calculated.[9]

4. Preparation of a Urea Electrode

Guilbault and Montalvo[10] have prepared a urea electrode by polymerizing urease in a polyacrylamide matrix on 100 micro dacron and nylon nets. These nets were placed over the Beckman 39137 cation selective electrode (which responds to NH_4^+ ion). The resulting "enzyme" electrode responds only to urea. The urea diffuses to the urease membrane where it is hydrolyzed to NH_4^+ ion. This NH_4^+ ion is monitored by the ammonium ion-selective electrode, the potential observed being proportional to the urea content of the sample in the range 1.0 to 30 mg of urea/100 ml of solution. This enzyme electrode appears to possess stability (the same electrode has been used for weeks with little change in potential readings or drift), sensitivity (as little as $10^{-4}\underline{M}$ urea is determinable) and specificity. Results are available to the analyst in less than 100 seconds after initiation of the test, and the electrode can be used for individual samples or in continuous operation.

5. Other Studies

Wieland, Determan and Buennig[11] also prepared insoluble enzymes in polyacrylamide gels. The

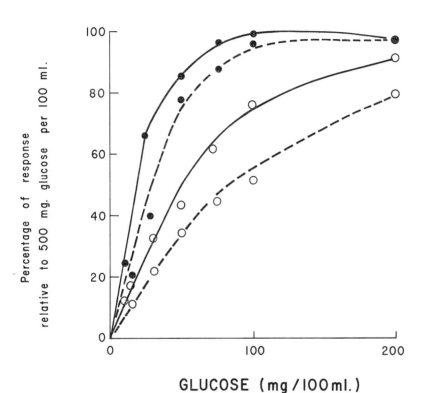

FIG. 10

Response of enzyme electrode with
different gels:—1,000 mg glucose
oxidase/100 ml of gel; ---, 100 mg
glucose oxidase/100 ml of gel; 0,
8 per cent gel concentration;0 , 19
percent gel concentration.(ref. 9).

enzymes alcohol dehydrogenase, trypsin and lactate
dehydrogenases were immobilized by physical entrapment using a procedure similar to that described
above.

D.   OTHER METHODS FOR IMMOBILIZING ENZYMES
1. General

A review on the preparation of insoluble enzymes
has been prepared by Chibata and Tosa.[12]   The
techniques of combining active enzymes with some
insoluble carrier either with covalent bonds, ionic
combination or physical adsorption are discussed.

Enzymes have been diazotized to cellulose
particles[13] and to polyaminostyrene beads.[14]
McLaren and Peterson[15], Nikolaev and Mardashev[16]
and Barnett and Bull[17] have successfully attempted
the physical entrapment of the enzymes asparaginase,
ribonuclease, and chymotrypsin by adsorption, absorption or ion exchange. Enzymes have also been immobilized on polytyrosyl polypeptides[18], on a collodion matrix[19] and encapsulated in semipermeable
micro-capsules made of synthetic polymers.[20]

Habeeb[21] manufactured water insoluble derivatives of trypsin using glutaraldehyde to conjugate
trypsin to aminoethyl cellulose. Weetall and Weliky[22]
have described the synthesis and continual operation
of a carboxymethylcellulose enzyme column, and the
manufacture of a similar enzyme paper preparation
which still retains its activity after 2 months
storage without refrigeration.[23]   Reese and
Mandels[24] described a method of obtaining an essentially continuous enzyme reaction on a two-phase
column utilizing partition chromatography. The

enzyme was retained as the stationary phase on a
column of the hydrophilic solid, cellulose. Enzymes
immobilized by binding to carboxymethyl cellulose
are available from Serevac (Maidenhead, England).

2. Covalent Bonding to Polymeric Lattices

Scientists of the Department of Biophysics of the
Weismann Institute (Reharoth, Israel) have pioneered
in the preparation of enzymes insolubilized by covalent bonding to polymeric lattices. These modified enzymes retain significant fractions of their
native activities while, according to initial studies,
certain other properties have in fact been
altered.[25-27] Three of the insolubilized co-
valently-bound enzymes (trypsin, chymotrypsin and
papain) are available as lyophilized powders from
Miles (Elkhart, Indiana).

a. Insolubilized Trypsin and Chymotrypsin. An
appropriate amount of trypsin or chymotrypsin is
added to a copolymer of maleic anhydride and ethylene, previously cross-linked with hexamethylene
diamine to decrease its water solubility. The reaction occurs in buffer solution overnight at $4^{\circ}C$.

By altering the ratio of enzyme to carrier, derivatives of differing characteristics are produced.[25]

Applications of insolubilized trypsin and chymotrypsin include:

1.   In amino acid sequence analysis - Improve
monitoring of proteolysis. A slower rate of proteolysis results and longer peptide fragments are produced by insolubilized trypsin and chymotrypsin.
Depending on the nature of the electrostatic interaction between the insoluble carrier and the protein
substrate, a smaller number of peptide bonds are
usually split.[28]

2.   The possibility of the isolation of specific
trypsin and chymotrypsin inhibitors - using columns
containing the insolubilized preparations.[29,30]

3.   In immunological separations - for separating
antibodies specific to trypsin and chymotrypsin from
sera.[31]

b.   Insolubilized Papain. The form provided is
prepared by coupling native papain to a water-
insoluble diazonium salt derived from a copolymer
of p-amino-DL-phenylalanine and L-leucine, the reaction occurring at 4°C. over a 20 hour period.[27,32]
The product is a stable, water-insoluble papain derivative retaining up to 70% of the original papain
activity on low molecular weight substrates and up
to 30% on high molecular weight substrates.

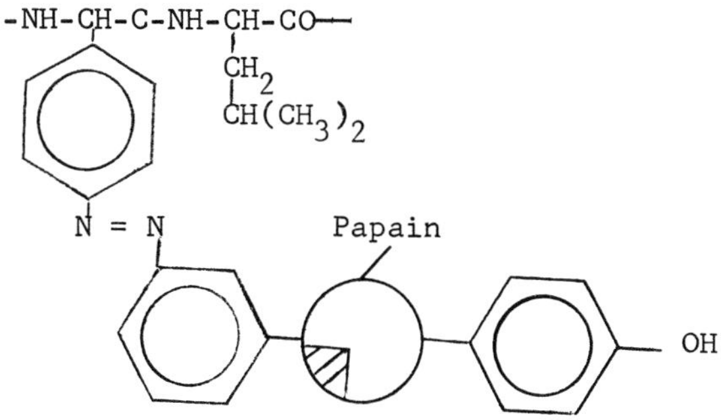

$-NH-CH-C-NH-CH-CO—$

CH_2

$CH(CH_3)_2$

$N = N$ Papain

OH

This insolubilized papain preparation has been used
to study the structure of rabbit γ-globulin. Because
this product is active in the hydrolysis of protein
in the absence of added reducing agents, it is possible
to differentiate protein fragments produced by
proteolysis from those produced by reduction.[32]

Katchalski[33] prepared water insoluble derivatives of papain by adsorption of papain chemical
derivatives on a collodion column. The acetyl-,
succinyl-, poly-L-ornithyl-, poly-γ-benzyl-L-glu-
tamyl-, water insoluble (maleic acid-ethylene)-,
and (4-amino-biphenyl-4'-N'-aminoethyl)-starch-
papain derivatives were prepared for investigation.

E. COMMERCIAL AVAILABILITY OF IMMOBILIZED ENZYMES

It is likely that immobilization methods will
soon be developed for all enzymes. Several companies
(Miles, Mann, Serevac) already offer some immobilized
enzymes and many more will probably be commercially

available soon. Polysciences (Harrington, **Pa.**)
supplies polymer substrates for enzyme immobilization as described by Katchalski and co-workers.

As the commercial availability of immobilized
enzymes increases, so will the number of analytical
applications.

F.   FUTURE APPLICATIONS

The immobilized enzyme will likely bring a new
future to enzymic analysis and to biochemistry in
general. Enzyme electrodes (transducers with
immobilized enzymes), similar to those described
for glucose by Hicks[8], would allow simple, direct,
continuous in vivo analysis of important body
chemicals. A glucose electrode, for example, would
permit a continuous analysis of blood glucose levels
in patients, or the analysis of glucose in blood or
urine samples in a hospital or clinical laboratory
in as simple a manner as a pH measurement. Similarly, implanted transducers using immobilized
enzymes could be used for patient therapy. The
uses of immobilized enzymes in synthesis and therapy
would be limitless.

REFERENCES

1. B. Vasta and V. Usdin, Melpar, Inc., Falls
    Church, Va., Final Report Contract No. DA 18-
    108-405-CML-828, Section 3.3.4, p. 3.102
    (Oct., 1963).

2. G. G. Guilbault, E. K. Bauman, D. N. Kramer and
    L. H. Goodson, Anal. Chem. 37, 1378 (1965).

3. G. G. Guilbault and J. Das, in preparation

4. G. G. Guilbault, M. Sadar, R. Glazer and C. Skou,
    Anal. Letters, 1, 333 (1968).

5. G. G. Guilbault and D. N. Kramer, Anal. Chem. 37, 1675 (1965).

6. P. Bernfeld and J. Wan, Science 142, 678 (1963).

7. G. P. Hicks and S. J. Updike, Anal. Chem. 38, 726 (1966).

8. S. J. Updike and G. P. Hicks, Science 158, 270 (1967).

9. Ibid., Nature 214, 986 (1967).

10. G. G. Guilbault and J. Montalvo, in preparation.

11. T. Wieland, H. Determan and K. Buennig, Z. Naturforsch. 21, 1003 (1966).

12. I. Chibata and T. Tosa, Tampakushitsu Kakuson Koso, 11, 23 (1966).

13. M. A. Mitz and L. J. Summaria, Nature 189, 576 (1961).

14. N. Grubhofer and L. Schleith, Naturwissenshcaften 40, 508 (1953).

15. A. D. McLaren and G. H. Peterson, Soil Sci. Soc. Am. Proc. 22, 239 (1958).

16. A. Nikolaev and S. R. Mardashev, Biokhimia 26, 565 (1962).

17. L. B. Barnett and H. B. Bull, Biochim. Biophys. Acta 36, 244 (1959).

18. A Bar-Eli and E. Katchalski, J. Biol. Chem. 238, 1690 (1963).

19. R. Goldman, H. I. Silman, S. Caplan, O. Kadern and E. Katchalski, Science 150, 758 (1965).

20. T. M. Chang, Science 146, 524 (1964).

21. A. Habeeb, Arch Biochem. Biophys. 119, 264 (1967).

22. H. Weetall and N. Weliky, Jet Propulsion Laboratory Space Programs Summary No. 37-26, 4, 160 (1965).

23. H. Weetall and N. Weliky, Anal Biochem. 14, 160 (1966).

24. E. T. Reese and M. Mandels, J. Am. Chem. Soc. 80, 4625 (1958).

25. Y. Levin, M. Pecht, L. Goldstein and E. Katchalski, Biochemistry 3, 1905 (1964).

26. L. Goldstein, Y. Levin and E. Katchalski, Biochemistry 3, 1913 (1964).

27. I. H. Silman, M. Albu-Weissenberg and E. Katchalski, Biopolymers 4, 441 (1966).

28. E. B. Ong, Y. Tsang and G. Perlmann, J. Biol. Chem. 241, 5661 (1966).

29. H. Fritz, H. Schult, M. Neudecker and E. Werle, Angew. Chemie International, 5, 735 (1966).

30. H. Fritz, H. Schult, M. Hutzel, M. Wiedemann and E. Werle, Hoppe-Seyler's Z. Physiol. Chemie 348, 308 (1967).

31. A. H. Sehon, International Symposium of Immunological Methods of Biological Standardization, Royaumont 1965, Symp. Series Immunobiol. Standards 4, 51, Karger, Basel/New York (1967).

32. J. J. Cebra, D. Givol, H. I. Silman and E. Katchalski, J. Biol. Chem. 236, 1720 (1961).

33. E. Katchalski, Technical Report AFOSR Grant 67-2025, June 30, 1967.

CHAPTER 8

USE OF AUTOMATION IN ENZYMIC ANALYSIS

A. GENERAL

Many of the experimental difficulties of using
enzymes in analysis by reaction rate methods could
be eliminated or lessened by the use of automation.
Ideally, all the steps in an enzymic procedure would
be automated: the addition of reagents, the measurement of the reaction rate, and the calculation of
results. Excellent reviews have been written by
Schwartz and Bodansky[1] and Blaedel and Hicks.[2]

In order to automate an enzyme procedure, generally
the rate of reaction must be calculated. From this
the amount of substance being analyzed in solution
can be determined. To do this, a pseudo first order
condition is established by adding excess quantities
of all reactants except the one to be assayed. The
measurement of the initial rate is made, and this is
proportional to the concentration of the substance
being determined (i.e., A): $A + B \xrightarrow{E} X + Y$.

At excess B and E, $\dfrac{dX}{dt} = k\left[A\right]$ (1)

where $[A]$ represents the concentration of A at time t
and k is a pseudo first order rate constant with
magnitude depending upon many factors including pH,
temperature, enzyme activity and concentrations of B
and E.

265

There are several possible methods to determine
the concentration A (or E) via the rate of reaction: 1) the initial slope method; 2) fixed concentration or variable time method; and 3) fixed
time method. All three of these methods can be
automated, and each will be discussed separately
below.

B. INITIAL SLOPE METHOD

In this method the change in the concentration
of a reactant, product or indicator substance is
plotted as a function of time by an automatic
recording of the rate curve. The initial slope
of this curve is obtained by extrapolation to time
zero and is related to the concentration of reactant or enzyme to be determined. Generally, the
change in some physiochemical parameter of the
reactant, product or inhibitor is measured (absorbance, pH, fluorescence, etc.). The initial slope
method is illustrated in Fig. 1, which shows typical
curves obtained in the lipase catalyzed conversion
of a non-fluorescent ester, dibutyrylfluorescein,
to the highly fluorescent fluorescein.[3] In these
curves the fluorescence is zero before lipase is
added. Upon addition of lipase, the fluorescence
increases due to production of fluorescein. The
initial rate of reaction is obtained by drawing a
straight line at the initial slope of the curve.
A plot of the change in fluorescence with time,
$\Delta F/\Delta t$, vs. concentration of enzyme should be linear,
allowing one to assay the enzyme concentration by
noting the rate of reaction (expressed as $\Delta F/\Delta t$).

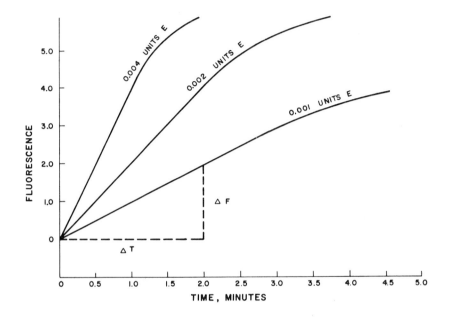

FIG. 1

Fluorescence - time curves for the hydrolysis
of dibutyrylfluorescein catalyzed by various
amounts of lipase.

Most spectrophotometric procedures can be easily
adapted to the same sort of partial automation,
since provision for recording at a single wavelength
is provided (Beckman DB or DK, Cary 15, etc.). The
Beckman DU can be modified for automatic recording of
reaction rate curves.[4] Equipment available from
Gilford, Inc. (Oberlin, Ohio) permits the simultaneous recording of the absorbance change in four
cuvets, using an automatic cuvet-positioning attachment.

Any hydrolytic enzyme reaction that liberates acid
or base as a product of the reaction can be followed
with a pH stat type approach. Jacobson[5] has published a review on the pH stat, its theory and appli-
cation to automatic recording of rate curves.

The automatic recording of reaction rates greatly
increases the quality and quantity of data measured
and eliminates the need for continuous attention;
however, only partial automation is realized, since
many manipulations are still necessary: the slopes
of the curves must be manually determined, and the
concentration calculated. For full automation,
generally a variable or constant time method is used.
This in effect means that method 1 becomes either
method 2 or method 3.

C. FIXED CONCENTRATION (VARIABLE TIME) METHODS

In this method the time that is required for the
concentration of a reactant to reach a set level is
recorded. Any property of this substance denoting
its concentration (e.g. its fluorescence, color)
could be used. Thus, the time required for a preset
fluorescence or absorbance level to be reached, would

be inversely proportional to the concentration, and a
plot of 1/t vs. concentration would be linear (Fig. 2).
This is predictable from basic kinetics.[6,7,8]
Integration of equation 1 yields

$$X = K[A]_o t \qquad (2)$$

$$[A]_o = K' \left(\frac{1}{t}\right) \qquad (3)$$

where K' is a constant which includes a constant,
specified concentration of substance at that point
in the reaction at time t. Thus, if the reciprocal
of the time required for the fluorescence to reach
a value of 2.0 (Fig. 1) is taken this is plotted vs.
the concentration of enzyme, a staight line calibration plot results (Fig. 2). Thus, one might determine the concentration of substrate A or enzyme E
by noting the time it takes to reach a fluorescence
value of 2.0.

In a typical automated procedure using the variable time method, the reaction is initiated by the
injection of enzyme. After a delay time (usually
about 30 seconds to allow the attainment of a steady
rate), a timer starts automatically. After a fixed
change in absorbance or fluorescence (measured by
the sensing unit) is reached, the timer automatically
shuts off and a reading is taken. The concentration
of substance present in the sample is inversely
proportional to the time required to reach a fixed
absorbance or fluorescence change. An analog to
digital converter and printer can be used to convert the time signal to a direct printout of concentration.

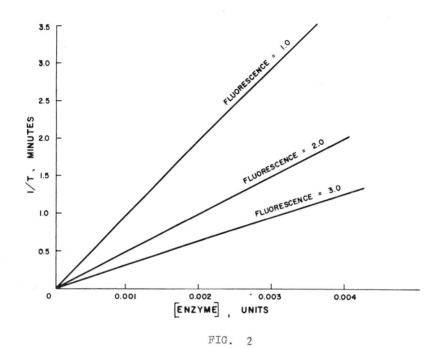

FIG. 2

Plot of 1/t vs enzyme concentration at three different
fluorescence values. (Data taken from Figure 1).

FIG. 2

Plot of 1/t vs. enzyme concentration at
three different fluorescence values.
(Data taken from Fig. 1).

In the fixed concentration procedure the time is measured between two precise points near the initiation of the reaction. One point is not zero time, since many reaction curves are not linear immediately after initiation due to mixing, temperature and stabilization effects. Generally 15-30 seconds are sufficient to establish a measurable rate.

Pardue and Malmstadt have developed automatic electrochemical methods for the determination of glucose oxidase and glucose based on the oxidation of iodide to iodine by hydrogen peroxide in the presence of molybdate as catalyst:

$$\text{Glucose} \xrightarrow{\text{Glucose Oxidase}} H_2O_2$$

$$H_2O_2 + I^- \xrightarrow{\text{MoO}_4} I_2$$

The iodine produced, whose rate of production is proportional to the rate of oxidation of glucose, is detected either potentiometrically[9,10,11] or amperometrically.[12,13] In either case, automatic control equipment provides a direct readout of the time required for a predetermined amount of iodine to be produced. The reciprocial of the time interval is proportional to the glucose oxidase activity or glucose concentration with relative standard deviations of about 2%.

Pardue[14] has extended the electrochemical techniques described to the assay of galactose and galactose oxidase. The hydrogen peroxide produced again reacts with iodide to form iodine, which is detected amperometrically. The reciprocal of the time interval required for a certain current to be produced is proportional to the enzyme and substrate

with a relative standard deviation of 2%. Toren[7]
has described a kinetics experiment for an analytical course using the enzymic determination of glucose
and both variable and fixed time methods.

Weinburg[15] has designed a completely automatic
instrument that stores 100 samples, adds reagents,
measures the time required for a preset absorbance
change to occur, records the data on a paper tape
and shuts off automatically. In the assay of ATPase
a pH indicator dye, phenol red, was used and the
time was measured for an absorbance change from
0.350 to 0.300 to occur

D. FIXED TIME METHOD

This method is similar to the fixed concentration
method. In this method the reaction is allowed to
proceed for a predetermined time internal, and after
this time has passed, the concentration of one of
the reactants (or products) is determined by some
physiochemical means (absorbance, fluorescence,
pH).

Thus in Fig. 1, one would measure the fluorescence produced by enzyme reaction at a set time,
say 1.0 minute. A calibration plot of fluorescence
(at t = 1.0 min) vs. concentration of enzyme would be
linear, thus allowing one to calculate the concentration of an unknown by a simple measurement of
its fluorescence at a preset time of 1.0 minute.
The time chosen should be one which would yield a
linear plot of F vs. concentration. Thus, if one
chose 1.0 min from Fig. 1, one would get a linear
curve (Fig. 3). But non-linear curves would result
if one chose either 1.5 or 2.0 minutes   This is
evident from Fig. 1.

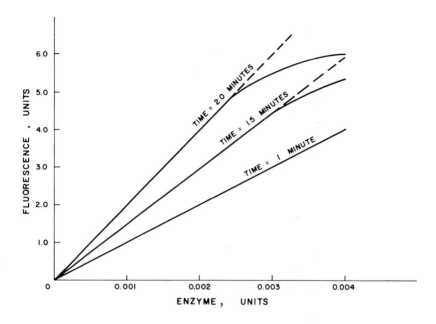

FIG. 3

Plot of fluorescence vs. enzyme concentration at three different times. (Data taken from Figure 1).

FIG. 3

Plot of fluorescence vs. enzyme concentration at three different times. (Data taken from Fig. 1).

J

Again an analog to digital converter and printer
can be used to convert the signal change to a direct
reading of the concentration of substance assayed.

Blaedel and Hicks have described automated assay
procedures for glucose[16] and lactate dehydrogenase
in blood serum[17] using fixed time methods.   Fig. 4
outlines the method for continuous assay of LDH.
The LDH reagent, containing all components of the
reaction except the enzyme, flows at a constant rate
to meet and mix with the sample stream containing the
enzyme. As the enzymic reaction occurs, the absorbance of the resulting stream changes continuously
as it flows away from the mixing point. The stream
passes through an upstream delay to overcome any
nonlinear induction effects that may exist, then to
the upstream cell, through an intercell delay and
finally into the downstream cell. Since constant
flow rates are used, the time interval between the
cells is fixed and the difference in absorbance
between the upstream and downstream cells is proportional to the rate of reaction. The absorbance
difference is measured with a sensitive differential
recording filter photometer, and is directly related
to the amount of LDH present.

In this analysis the extent of reaction is kept
small, and the measured rate is virtually an initial
rate. Very small changes in reactant concentrations
occur. Thus, a direct proportionality between the
measured absorbance change and the concentration of
the substance assayed is achieved. Also, the measurement of the difference in absorbance eliminates many
errors due to the presence of absorbing, non-reactive
impurities in the sample. The system is simple, easy

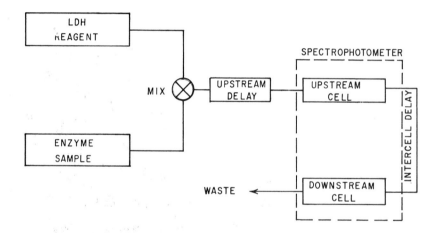

FIG. 4

Outline of continuous measurement of
LDH (Redrawn from reference 17).

to build, and requires little manipulation of sample
or reagents. A chart record of the results (Fig. 5)
allows a direct readout of the concentration of
enzyme present with no calculations necessary.
Blaedel and Hicks[16] assayed glucose in aqueous
solutions up to 60 ppm with a standard deviation of
1 ppm. Up to 15 samples per hour were analyzed,
with a readout time of 4 minutes per sample. For
LDH, Blaedel and Hicks[17] analyzed serum samples
(0.2 ml) at a rate of 20 per hour. A direct readout of LDH units could be accomplished (Fig. 5)with
an error of about 20 LDH units. The absorbance-
change indicator system used for LDH[14] was 2,6-
dichloroindophenol-NAD-diaphorase:

$$\text{L-Lactate} + \text{NAD} \xrightarrow{\text{LDH}} \text{pyruvate} + \text{NADH} + \text{H}^+$$

$$\text{NADH} + \text{H}^+ + \text{Dye}_{ox} \xrightarrow{\text{Diaphorase}} \text{NAD} + \text{dye}_{red}$$

$$\text{(blue)} \qquad\qquad\qquad \text{(colorless)}$$

and that for glucose[16] was peroxidase-o-dianisidine:

$$\text{Glucose} + O_2 \xrightarrow{\text{Glucose Oxidase}} H_2O_2$$

$$H_2O_2 + \text{o-dianisidine} \xrightarrow{\text{Peroxidase}} \text{Red color}$$

Toren[7] and Pardue[8] have described kinetic
experiments using the enzymic glucose determination
and fixed time procedures. Amperometric[8] and
absorbance[7] methods were used.

Blaedel and Olson[18] have developed an automated
system for glucose similar to that described above,
except for an electrochemical readout. A differen-

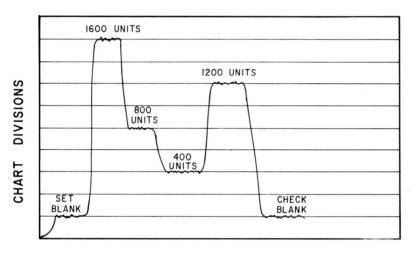

TIME

FIG. 5

Record of LDH samples by fixed time
procedure (Redrawn from reference 17).

tial amperometric procedure is used, based upon the
continuous measurement of the rate of the glucose
oxidase reaction in a flowing system. The H_2O_2
produced oxidized ferrocyanide to ferricyanide,
which is measured with a tubular platinum electrode:

$$H_2O_2 + 2Fe(CN)_6^{-4} \xrightarrow{2H^+} 2Fe(CN)_6^{-3} + 2H_2O$$

The sample (containing glucose) and glucose oxidase-
ferrocyanide reagent solution are mixed and the
resulting solutions flow through a delay line to
permit complete mixing and to eliminate induction
effects. The stream then flows through an upstream
tubular platinum electrode (TPEu), an interelectrode
delay line, then through a downstream electrode (TPEd)
(Fig. 6). The difference in concentration of electroactive substance at the two electrodes is found by a
differential amperometric measurement at constant
potential. Assuming a constant flow rate and a
fixed distance between electrodes, the current
measured is proportional to the difference in concentration of electroactive substance at the 2 electrodes, and therefore is proportional to the reaction
rate and to the glucose concentration.

The recorded response is linear with glucose concentration up to 100 ppm (Fig. 7) and up to 20
samples per hour can be run.

All the methods described above can easily be
provided with direct digital readout. The analog
response is simply converted to a digital signal
which can be printed out in concentration units.
Malmstadt and Piepmeier have developed an automatic
pH stat with digital readout for quantitative
enzyme determinations.[19] A stability of $\pm$ 0.002 pH

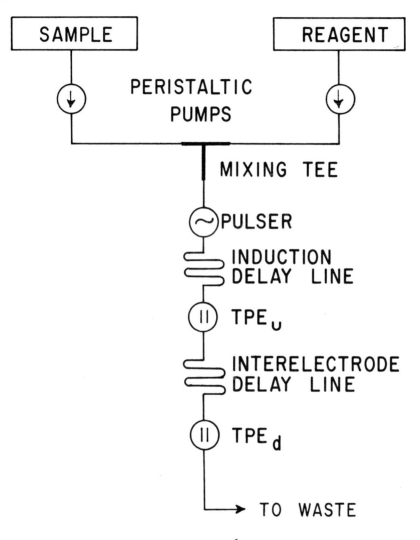

FIG. 6

Outline of method of continuous analysis
by amperometric measurement of reaction
rate (ref. 18).

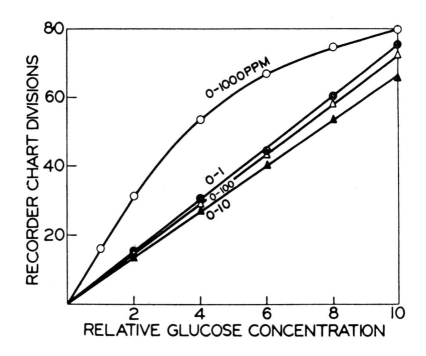

FIG. 7

Working curves for standard glucose
samples (ref. 18).

Horizontal axis refers to relative
glucose concentrations in the 0-1,
0-10, 0-100, and 0-1000 ppm range.

units is reported. Pardue[20] has designed automatic control equipment to provide digital readout
of concentration data in reaction rate methods.
Pardue has also described an automatic method for
measuring the slopes of rate curves.[21]

The Analytical Instrument Division of American
Optical Corporation (Richmond, California) markets
an instrument that completely automates routine
enzymic analysis procedures. The instrument, the
Robot Chemist, can handle 120 samples per hour and
carries out all the steps of the enzymic procedure:
addition of sample and reagent, mixing, incubation
of the reaction, spectrophotometric measurement,
calculation of results, and direct printout. The
instrument uses the fixed time procedure, measuring
the absorbance change resulting from the enzymic
reaction, which is then proportional to the enzyme
or substrate to be assayed. A schematic diagram of
the Robot Chemist specimen handling capability is
pictured in Fig. 8. In an analysis, a measured
aliquot of the sample is picked up from the rack of
samples at A. This aliquot is transferred to a
reaction tube in the process turntable at B, followed
by a pre-determined volume of reagent previously
picked up at C. One, two or four aliquots, followed
by a single or alternating reagents, may be programmed for pickup from each specimen and delivered
to adjacent reaction tubes in the turntable. The
turntable, containing reaction tubes immersed in a
temperature-controlled incubation bath, advances one
position every 30 seconds. Additional reagents may
be added at any position (such as D) during the
incubation period. When the incubation is complete,
the reacted mixture is transferred (at point E) to

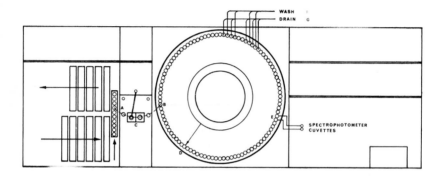

FIG. 8

Robot Chemist

Schematic diagram of speciment handling
capability. (Compliments of American
Optical Corporation).

the cuvette of the spectrophotometer for measurement and digital printout of concentration units.
The emptied reaction tubes are washed at positions
F and drained at positions G. Thus all that is
required is that a technician load the serum samples
to be analyzed, start the machine and read the
results. The Robot Chemist console is pictured in
Fig. 9 .

Bausch and Lomb markets an instrument for
unattended automatic enzymic analysis. The Zymat
340 (Fig. 10) does all the pipetting, measuring,
mixing, stirring and heat controlling normally done
by the lab technician. As many as 47 samples can be
handled in each loading, all highly precise answers
are printed out directly in International Enzyme
Units with identifying serial numbers. The instrument is intended for lactate dehydrogenase (LDH),
glutamate oxalate transaminase (SGOT) and glutamate
pyruvate transaminase (SGPT) analysis, but should
be adaptable for other determinations.[22] The
instrument uses a spectrophotometer monitoring of
the enzymic reaction.

The Du Pont Company (Instrument Products Division, Wilmington) has developed an instrument, the
Automatic Clinical Analyzer (ACA), that is designed
to reduce to a minimum the time between sampling
and transmittal of precise laboratory data. A
separate pack is provided for each test performed
on a sample in an ACA. Each pack contains both
the test name for convenient operator identification
and binary code to instruct the instrument. The
lab technician programs the analyzer by inserting
the appropriate pack or packs behind each sample
cup in the ACA input tray.

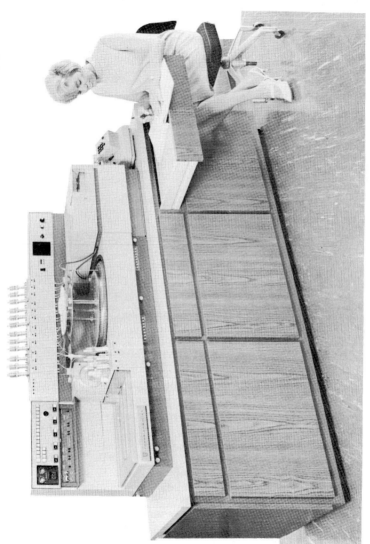

FIG. 9

Robot Chemist

(Compliments of American Optical Corp.)

FIG. 10

Zymat 340

(Compliments of Bausch and Lomb)

The analyzer automatically injects the exact
amount of sample and diluent into each pack in
succession, mixes the reagents, waits a preset
amount of time, forms a precise optical cell within
the transparent pack walls, and measures the reaction photometrically. These operations are controlled and monitored by a built-in, solid-state,
special-purpose computer, and are performed under
precisely regulated conditions within the instrument. The computer calculates the concentration
value for each test and prints out the results on
a separate report sheet for each sample. This
report contains all the test results on that sample
along with the patient identification. The used
test packs are discarded automatically into a waste
container.

The instrument has a coefficient of variation
of 1-3%, and the first result is obtained in less
than 7 minutes after sample injection. In continuous operation successive test results are obtained
every 35-70 seconds.

Some of the enzyme tests for which the ACA is
programmed include glucose (using glucose oxidase-
peroxidase), urea (with urease and glutamate dehydrogenase), alkaline phosphatase (with p-nitrophenylphosphate), pseudocholinesterase, lactate dehydrogenase, hydroxybutyric dehydrogenase and aspartate
aminotransferase.

The assay of urea nitrogen (BUN) illustrates
the test operation. Urease specifically hydrolyzes
urea to form ammonia and carbon dioxide. This
ammonia is utilized by the enzyme glutamic dehydrogenase (GDH) to aminate α-keto-glutarate. Since

NADH is required for the amination, the reaction
rate is measured by observing the decrease in absorbance at 340 nm (mμ).

$$\text{Urea} + H_2O \xrightarrow{\text{Urease}} 2\ NH_3 + CO_2$$

$$NH_3 + NADH + H^+ + \alpha\text{-ketoglutarate} \xrightarrow{\text{GDH}} NAD +$$

(absorbs at   L-glutamate
340 mμ)

Du Pont expects to start marketing the ACA
following additional field evaluations and estimates
the price to be $65,000.00.

Another continuous system that carries out all
the manipulations is the AutoAnalyzer, originally
described by Skeggs[23] and available from Technicon, Inc. (Chauncey, New York). The AutoAnalyzer
uses continuously flowing streams metered propor-
tionally by a single multichannel peristaltic pump,
and uses either a colorimetric or fluorometric readout of concentration changes at a fixed time. In
AutoAnalysis chemical reactions take place in continuously flowing, air-segmented systems.

A typical schematic diagram of an AutoAnalyzer
system is pictured in Fig. 11 along with the actual
components used. Briefly, operation is as follows:
the samples to be analyzed are loaded into the cups
on the sampler, and a multiple-channel proportioning
pump, operating continuously, moves the samples, one
following another, and a number of streams of reagents, into the system. Sample and reagents are
brought together under controlled conditions, causing
a chemical reaction and color development. Color

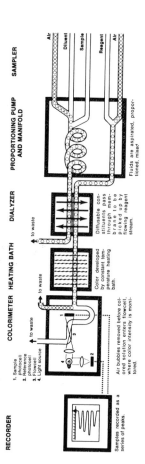

Typical Single Channel Flow Schematic

FIG. 11

intensity of the analytical stream is measured in a
colorimeter and the results of any analysis are presented as a series of peaks on a recorder chart
(Fig. 12).

Fundamental to AutoAnalyzer techniques is the
exposure of a known standard to exactly the same
reaction steps as the unknown samples. The concentrations of the unknowns are continuously plotted
against the known concentrations. Hence reactions
need not be carried to completion as in conventional
procedures.

A variety of analyses can be performed with the
AutoAnalyzer using detection methods that include
colorimetry, spectrophotometry, flame photometry,
fluorometry and atomic absorption spectrometry. A
bibliography of 1825 papers describing automated
analysis with the AutoAnalyzer in the last 10 years
is available from Technicon.[24] Guilbault in his
review articles on the Use of Enzymes in Analytical
Chemistry[25, 26] has likewise listed many enzymic
analyses that can easily be performed using the
AutoAnalyzer. Typical analyses include: acid and
alkaline phosphatase, amylase, cholinesterase,
glucose oxidase, LDH, SGOT, SGPT, lactic acid, glucose, uric acid and triglycerides, to name but a few.

As methods have become available which take advantage of the inherent high sensitivity and specificity of fluorometry, researchers have adapted these
to the AutoAnalyzer. In a number of recent applications the fluorescence of NADH, when excited at 340 mμ,
provides a common method for the measurement of all
enzyme systems that involve NAD or NADP. Methods have
been developed for the enzymes LDH[27], SGOT and
SGPT[28] using a fluorometric readout. Technicon has

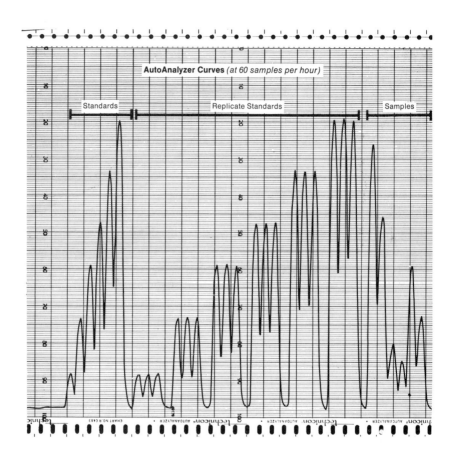

FIG. 12

developed a new, highly sensitive, very stable
fluorometer for use with these new methods.   Built
to form an integral part of automated continuous
flow systems, the fluorometric AutoAnalyzer can
handle up to 60 samples per hour.

Aminco and Turner Instrument Companies also
market fluorometers that can be used with the Auto-
Analyzer in flow systems.

Another recent development in automation is a
multichannel analyzer developed by Skeggs[29] which
determines 12 substances in a single 2 ml sample of
blood.   Technicon has marketed this concept in the
SMA 12/60 Analyzer which can run 60 such samples per
hour.   The time from aspiration of a given sample to
finished chart is only 9 minutes.   Results are automatically recorded on a precalibrated strip chart,
the Serum Chemistry Graph (Fig. 13).   The final
product is a comprehensive chemical profile of each
patient available quickly and at less cost than the
few tests he now receives.   The enzymes SGOT, alkaline
phosphatase and LDH, and glucose and urea are determined by enzymic methods, in addition to other important biochemical substances determined non-enzymatic-
ally.   Fig. 13 shows a serum chemistry graph of a
patient with severe diabetes mellitus.   The black
line, which indicates the levels of each substance
determined by the AutoAnalyzer, shows elevated SGOT,
alkaline phosphatase, glucose and billirubin levels
(shaded areas represent the normal levels).

The final step in automating the clinical lab
is represented by the Technicon On Line computer
system.   It will monitor, calculate, store and
report results, printed out in concentration units
with associated sample identification numbers   from

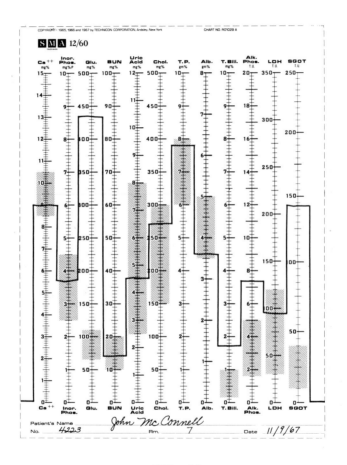

FIG. 13

Serum chemistry graph of a patient with severe
diabetes mellitus. The following biochemical
abnormalities can be seen: elevated SGOT,
alkaline phosphatase, glucose and billirubin.

any combination of AutoAnalyzer and/or other instrument systems. This unit will be marketed by Technicon
in the near future.

In this chapter an attempt has been made to review briefly what has been done recently in the automation of enzymic analysis. In the next decade an
upsurge in attempt towards complete automation will
undoubtedly result. Many new instruments will be
marketed, and the present instruments will unques-
tionably improve. The clinical or hospital lab of
the 1970's will be one that performs many enzymic
analysis routinely, all with complete automation and
with computer assisted storage and evaluation of data.

REFERENCES

1. M. K. Schwartz and O. Bodansky, "Methods of
   Biochemical Analysis," Vol. XI, (D. Glick, ed.),
   pp. 211-246, Interscience, New York, 1963.

2. W. J. Blaedel and G. P. Hicks, "Advances in
   Analytical Chemistry and Instrumentation,"
   Vol. 3, (C. N. Reilley, ed.), pp. 105-140,
   Interscience, New York, 1964.

3. G. G. Guilbault and D. N. Kramer, Anal. Chem.
   36, 409 (1964).

4. W. A. Wood and S. R. Gilford, Anal. Biochem.
   2, 589 (1961).

5. C. F. Jacobson, "Methods of Biochemical Analysis," Vol. 4, (D. Glick, ed.), p. 771, Interscience, New York, 1957.

6. W. Moore, Physical Chemistry, pp. 528-537,
   Prentice Hall, New Jersey (1955).

7. C. Toren, J. Chem. Ed. 44, 172 (1967).

8. H. Pardue, M. Burk and D. O. Jones, J. Chem.
   Ed. 44, 684 (1967).

9. H. V. Malmstadt and H. L. Pardue, Anal. Chem.
   33, 1040 (1961).

10. Ibid., Clin. Chem 8, 806 (1962).

11. H. Pardue, R. Simon and H. Malmstadt, Anal. Chem. $\underline{36}$, 735 (1964).

12. H. Pardue, Anal. Chem. $\underline{35}$, 1240 (1963).

13. H. Pardue and K. Simon, Anal. Biochem. $\underline{9}$, 204 (1964).

14. H. Pardue and C. Frings, J. Electroanal. Chem. $\underline{7}$, 398 (1964).

15. D. I. Weinburg, IRE Intern. Conv. Record $\underline{8}$, Part 9, 88 (1960).

16. W. J. Blaedel and G. P. Hicks, Anal. Chem. $\underline{34}$, 388 (1962).

17. Ibid., Anal. Biochem. $\underline{4}$, 476 (1962).

18. W. J. Blaedel and C. Olson, Anal. Chem. $\underline{36}$, 343 (1964).

19. H V. Malmstadt and E. Piepmeier, Anal. Chem. $\underline{37}$, 34 (1965).

20. H. Pardue, C. Frings and C. J. Delaney, Anal. Chem. $\underline{37}$, 1426 (1965).

21. H. Pardue, Anal. Chem. $\underline{36}$, 633 (1964).

22. Bausch and Lomb Catalog 34-6016, Zymat 340, Bausch and Lomb, Rochester, New York

23. L. T. Skeggs, Am. J. Clin. Pathol. $\underline{28}$, 311 (1957).

24. Technicon AutoAnalyzer Bibliography 1957/1967, Technicon Corporation, Ardsley, New York.

25. G. G. Guilbault, Anal. Chem. $\underline{38}$, 527R (1966).

26. G. G. Guilbault, Anal. Chem. $\underline{40}$, 459R (1968).

27. L. Brooks and H. G. Olken, Clin. Chem. $\underline{11}$, 748 (1965).

28. J. B. Levine and J. B. Hill, Automation in Analytical Chemistry, Technicon Symposium, New York, Mediad, Inc., p. 569, 1965.

29 L. T. Skeggs, Abstract of Papers, Winter ACS Meeting, Phoenix, Arizona, 1966.

APPENDIX

Sources of Enzymes

Sources of enzymes are taken from manufacturer's catalogs available through September, 1968. The listing is intended to be complete. Complete addresses are listed at the end of Appendix I.

ENZYMES	SOURCES
1. Acetate kinase, _E. coli_	Boehringer
2. Acylase, Bacterial	Mann, P-L Biochemicals
hog kidney	BDH, Calbiochem, General Biochem, Mann, NBC, Pentex, Sigma
3. Adenosine Deaminase,	
calf intestine	Boehringer, Calbiochem, Sigma
4. 5'-Adenylic Acid Deaminase	Sigma
5. Alcohol Dehydrogenase,	
horse liver	Boehringer, Calbiochem, Mann, Merck, NBC, P-L Biochemicals, Seravac, Pierce, Sigma, Worthington
yeast	BDH, Boehringer, Calbiochem, Mann, Mycofarm-Delft, NBC, Pierce, P-L Biochemicals, Seravac, Sigma, Worthington
6. Aldolase, rabbit muscle	Boehringer, Calbiochem, Mann, NBC, Pierce, Sigma, Worthington
7. Amino Acid Oxidase, D-, hog kidney	BDH, Boehringer, Calbiochem, General Biochem, Mann, Miles, NBC, Pentex, Pierce, P-L Bio-

	chemicals, Seravac, Sigma, Worthington
L-, venom	Boehringer, Calbiochem, Mann, NBC, Pierce, Sigma, Worthington

8. γ-Aminobutyric Trans-
    aminase-Succinic
    Semialdehyde De-
    hydrogenase
	Calbiochem

9. Aminopeptidase, B. subtillis	Miles, P-L Biochem- icals, Sigma
10. Amylase, α-, bacterial	BDH, Calbiochem, Mann, Merck, Pierce, Sigma, Wallerstein
fungal	Calbiochem, Sigma, Wallerstein
malt	Pfanstiehl, Sigma, Wallerstein
pig pancreas	BDH, Calbiochem, General Biochem, Mann, Merck, NBC, Sigma, Worthington
β-, barley	BDH, Mann, Miles, Pierce, Sigma, Waller-stein
sweet potato	BDH, Calbiochem, General Biochem, Mann, Merck, NBC, Sigma, Worthington
11. Amyloglucosidase, mold	Merck, Sigma, Waller-stein
12. Anthocyanase, A. niger	Miles
13. Apyrase, potato	Sigma
14. Arginase, beef liver	Mann, NBC, Pierce, Sigma, Worthington
15. Arginine Decarboxylase E. coli	General Biochem, Mann, NBC, Sigma, Worthington

16. Aromatase, human placenta   Mann

17. Asparaginase, E. coli   Mann, Schwarz, Sigma,
                                  Worthington

18. Carbonic Anhydrase,
    bovine erythrocyte   BDH, Calbiochem, Mann,
                                 NBC, Pentex, Pierce,
                                 Schwarz, Seravac,
                                 Sigma

19. Carboxypeptidase,
    pancreas   Calbiochem, General
                                 Biochem, Mann, NBC,
                                 Pentex, Pierce, Schwarz,
                                 Sigma

20. Carnitine acetyltrans-   Boehringer
    ferase, pigeon

21. Catalase, beef liver   BDH, Boehringer,
                                 Calbiochem, Cudahy,
                                 General Biochem, Mann,
                                 NBC, Pentex, Pierce,
                                 Schwarz, Sigma, Wor-
                                 thington

    fungal   Calbiochem, Merck,
                                 NBC, P-L Biochemicals

22. Cellulase, A. niger   Calbiochem, Mann,
                                 Merck, NBC, Pierce,
                                 Sigma, Schwarz, Waller-
                                 stein, Worthington

    T. viride   Miles, Schwarz, Sigma,
                                 Worthington

23. Chitinase, Strep   Mann, NBC, Worthing-
    griseus   ton

24. Cholinesterase, Acetyl,
    bovine erythrocyte   Mann, NBC, Pierce, Sigma
    eel   Mann, Worthington
    Butyryl, horse serum   Calbiochem, Mann, NBC,
                                 Pierce, Schwarz, Sigma,
                                 Worthington

25. Chymotrypsin, α-, bovine pancreas

BDH, Boehringer, Cal-biochem, General Bio-chem, Mann, Merck, Miles, NBC, Pentex, Pierce, P-L Biochem-icals, Schwarz, Seravac, Sigma, Worthington

β-, bovine pancreas

Calbiochem, General Biochem, Mann, Merck, NBC, Pentex, Pierce, P-L Biochemicals, Schwarz, Seravac, Sigma, Worthington

γ-, bovine pancreas

Calbiochem, Mann, Merck, NBC, Pentex, Pierce, P-L Biochem-icals, Schwarz, Sigma, Seravac, Worthington

δ-, bovine pancreas

BDH, Mann, NBC, Pierce, Schwarz, Seravac, Wor-thington

26. Chymotrypsinogen, pancreas

BDH, Boehringer, Cal-biochem, Mann, NBC, Pentex, Pierce, P-L Biochemicals, Schwarz, Seravac, Sigma, Wor-thington

27. Citrate Lyase, Entero-bacter aerogenes

Boehringer

28. Citrate synthase, pig heart

Boehringer, Calbiochem, Sigma

29. Collagenase, bacterial

Calbiochem, General Biochem, Mann, Miles, NBC, Pierce, Schwarz, Sigma, Worthington

30. Creatine phosphokinase, rabbit muscle

Boehringer, Calbiochem, Mann, NBC, Schwarz, Sigma, Worthington

31. Cytochrome c Reductase, pig heart

Calbiochem, Mann, NBC, Seravac, Sigma

32. Deoxyribonuclease
        (DNase),
    bovine plasma   BDH, Calbiochem,
                             General Biochem, Mann,
                             Miles, Pentex, Pierce,
                             Seravac, Sigma, Wor-
                             thington

    bovine spleen   Calbiochem, Mann,
                             Miles, Pierce, Schwarz,
                             Seravac, Sigma, Wor-
                             thington

    hog spleen   Schwarz
    micrococcal   Miles

33. Diamine Oxidase, hog   Calbiochem, NBC,
        kidney   Pentex, Sigma

34. Diaphorase, Cl. kluy-   Mann, NBC, Schwarz,
        verii   Sigma, Worthington
    pig heart   Boehringer, Calbiochem,
                             General Biochem, Mann,
                             P-L Biochemicals,
                             Seravac, Sigma

35. Diastase (see Amylase)

36. Dioldehydrase   Calbiochem

37. DNA polymerase, M. lyso- Miles, Sigma
        deikticus
    E. coli   General Biochem (Bio-
                             polymers Div.), Wor-
                             thington

38. DPN ase (see NAD-Nucleo-
        sidase)

39. DPN kinase   Sigma

40. Elastase, bacterial   Calbiochem, General
                             Biochem, Mann

    fermentation   Mann
    pancreatic   Calbiochem, General
                             Biochem, Mann, NBC,
                             Pentex, Pierce,
                             Schwarz, Seravac,
                             Worthington

41. Enolase, rabbit muscle   Boehringer, Calbiochem, Mann, NBC, Pierce, Sigma

    yeast   Sigma

42. Enterokinase   Pentex

43. Ficin, fig latex   Calbiochem, General Biochem, Mann, Merck, NBC, Pentex, Schwarz, Seravac, Sigma

44. Fructose-6-Phosphate kinase, rabbit muscle   Boehringer, Calbiochem, Sigma

    yeast   Boehringer

45. Fumurase, pig heart   Boehringer, Calbiochem, Mann, NBC, Pierce, Sigma

46. Galactose dehydrogenase,

    Ps. fluorescens   Boehringer

47. Galactose oxidase, D. dendroides   General Biochem, Mann, NBC, Schwarz, Worthington

    P. circinatus   Miles, Pierce

48. ß-Galactosidase, bovine liver   NBC, Sigma

    fungal   Wallerstein

    yeast   BDH, NBC, Pierce

49. Glucose dehydrogenase,

    calf liver   Sigma

50. Glucose oxidase, A. niger   Mann, Miles, NBC, Pierce, Pfanstiehl, P-L Biochemicals, Schwarz, Sigma, Worthington

                 BDH, Boehringer, Calbiochem, General Biochem, Mann, Merck, Pierce

51. Glucose-6-phosphate dehydrogenase, yeast — BDH, Boehringer, Calbiochem, Mann, NBC, Pierce, P-L Biochemicals, Sigma

52. β-Glucosidase, almonds — BDH, Calbiochem, General Biochem, Pierce, P-L Biochemicals, Schwarz, Seravac, Sigma, Worthington

53. β-Glucuronidase, bovine liver — Calbiochem, Mann, Schwarz, Sigma, Worthington

    helix pomatia — Boehringer, Calbiochem, Sigma

    mollusc — BDH, Calbiochem, Mann, Pierce, P-L Biochemicals, Seravac, Sigma

54. L-Glutamic Decarboxylase, Cl. welchii — NBC, Schwarz, Sigma, Worthington

    E. coli — Mann, NBC, Pierce, Schwarz, Sigma, Worthington

55. L-Glutamic Dehydrogenase, bovine liver — Boehringer, Calbiochem, Mann, NBC, Sigma

56. Glutamic-oxalacetic transaminase, pig heart — Boehringer, Calbiochem, Mann, NBC, Worthington

57. Glutamic-pyruvate transaminase pig heart — Boehringer, Calbiochem, Mann

58. Glutaminase, E. coli — Mann, NBC, P-L Biochemicals, Schwarz, Sigma, Worthington

59. Glutamine Synthetase, sheep brain — Mann, NBC, P-L Biochemicals

60. Glutamine Decarboxylase, E. coli — Sigma

61. Glutathione Reductase,   Boehringer, Calbiochem,
    yeast   Mann, NBC, Sigma

62. Glyceraldehyde-3-phos-   Boehringer, Calbiochem,
    phate dehydroge-   NBC, Schwarz, Sigma,
    nase, rabbit   Worthington
    muscle

    yeast   NBC, Worthington

63. Glycerokinase, Candida   Boehringer, Calbiochem,
    mycoderma   NBC, Sigma

64. Glycerol dehydrogenase,   Mann, NBC, Pierce,
    Enterobacter aerogenes   Schwarz, Sigma, Wor-
                                    thington

65. Glycerol-1-phosphate   Boehringer, Calbiochem,
    dehydrogenase,   Mann, NBC, Sigma
    rabbit muscle

66. Glycerol-1-phosphate   Boehringer, Calbiochem,
    dehydrogenase.   Mann, Sigma
    Triosephosphate
    Isomerase, rabbit
    muscle

67. Glyoxalase, yeast   Boehringer, Calbiochem,
                                    NBC, Sigma

68. Glyoxalate Reductase,   Boehringer, Calbiochem,
    spinach   Pierce

69. Guanase, rabbit liver   Boehringer, Calbiochem,
                                    Mann, Sigma

70. Hemicellulase, mold   Mann, NBC, Pierce,
                                    Sigma

71. Hesperidinase, A. niger   Miles

72. Hexokinase, yeast   Boehringer, Calbiochem,
                                   General Biochem, Mann,
                                   Pierce, P-L Biochem-
                                   icals, Schwarz, Seravac,
                                   Sigma

73. Histidase, Ps. fluorescens   Mann, NBC, Sigma,
                                    Worthington

74. Histidine decarboxylase,   Mann, NBC, Pierce,
    Cl. welchii   Schwarz, Sigma, Wor-
                                thington

75. Hyaluronidase, bee venom   Sigma

    bovine testes   Calbiochem, Cudahy,
                                General Biochem, Mann,
                                NBC, Pentex, Pierce,
                                Schwarz, Seravac,
                                Sigma, Worthington

    ovine testes   BDH, Calbiochem,
                                Mann, Pierce, Sigma

76. Hydrolase powder, A.   Boehringer
    oryzae

77. Hydroxyacyl-Co A dehydro-   Boehringer, Calbiochem,
    genase, pig heart   Sigma

78. β-Hydroxybutyrate dehy-   Boehringer, Calbiochem,
    drogenase, Rhodo-   Sigma
    pseudomonas
    sphaeroides

79. β-Hydroxysteroid dehydro-   NBC, Schwarz, Sigma,
    genase, Ps. testo-   Worthington
    steroni

    Streptomyces hydro-   Boehringer, Calbiochem,
    genans   Schwarz, Sigma, Wor-
                                 thington

80. Inorganic pyrophospha-   Calbiochem
    tase, A. oryzae

    yeast   NBC, Sigma, Worthing-
                                 ton

81. Invertase, yeast   BDH, Calbiochem, Mann,
                                 Miles, NBC, Pfanstiehl,
                                 Sigma, Wallerstein

82. Isocitrate dehydrogenase,   Boehringer, Calbiochem,
    pig heart   Mann, NBC, Sigma

83. Lactic dehydrogenase,   Boehringer, Calbiochem,
     bovine heart   Mann, Miles, NBC,
                               Pierce,Schwarz, Seravac
                               Sigma, Worthington

     chicken heart   Mann, P-L Biochem-
                               icals

     pig heart   Boehringer, Sigma

     rabbit muscle   BDH, Boehringer,
                               Calbiochem, Mann, NBC,
                               Pierce, Schwarz, Sigma,
                               Worthington

     yeast   NBC, Schwarz, Sigma,
                               Worthington

84. Lecithinase (Phospho-
         lipase)
         A-, Snake   Boehringer, Calbiochem,
                               Pierce, Sigma

         B-, bovine pancreas   Pierce

         C-, Cl. perfringens   NBC, Schwarz, Wor-
                               thington

         C-, Cl. welchii   Calbiochem, General
                               Biochem, Mann, Pierce,
                               Sigma

         D-, cabbage   BDH, Boehringer, Cal-
                               biochem, General Bio-
                               chem, Mann, Pierce,
                               Sigma

85. Leucine aminopeptidase,   BDH, Boehringer, Mann,
     hog kidney   Miles, NBC, Pentex,
                               Pierce, P-L Biochem-
                               icals, Schwarz, Seravac,
                               Sigma, Worthington

86. Lipase, calf gland   Mann

     microbial   Mann, Miles

     pig pancreas   Calbiochem, Cudahy,
                               General Biochem, Mann,
                               Merck, NBC, Pfanstiehl,
                               Pierce, Schwarz, Sigma,
                               Worthington

     wheat germ   BDH, Calbiochem, Mann,
                               NBC, Pentex, Pierce,
                               Schwarz, Sigma, Wor-
                               thington

87. Lipoxidase, soybean

Mann, NBC, Pierce,
P-L Biochemicals,
Schwarz, Seravac,
Sigma, Worthington

88. Luciferase, _Photobacte-_
       rium fischeri

NBC, Schwarz, Sigma,
Worthington

89. Lysine decarboxylase,
       B. cadaveris

Calbiochem, Mann, NBC,
Pierce, Schwarz, Sigma,
Worthington

    E. coli

Sigma

90. Lysozyme (Muramidase),
       egg

BDH, Boehringer, Cal-
biochem, General Bio-
chem, Mann, Miles,
NBC, Pentex, Pierce,
P-L Biochemicals,
Schwarz, Seravac,
Sigma, Worthington

91. Malic dehydrogenase,
       pig heart

BDH, Boehringer, Cal-
biochem, Mann, Miles,
NBC, Pierce, P-L Bio-
chemicals, Schwarz,
Seravac, Sigma, Wor-
thington

92. Maltase, fungal

Sigma

93. Mylase, _A. oryzae_

Mann, NBC, Pierce

94. Myokinase, rabbit muscle

Boehringer, Calbiochem,
General Biochem, Mann,
NBC, Sigma

95. NAD-Nucleosidase
       (DPNase), calf
       spleen

Boehringer

    microccal

Mann, NBC, Sigma,
Worthington

    pig brain

Boehringer

96. Nariginase, _A. niger_

Miles

K

97. Neuraminidase, Cl. per-   NBC, Sigma, Schwarz
       fingens

       V. cholerae   BDH, Calbiochem,
                                  General Biochem, Mann,
                                  Pierce

       virus   Calbiochem, General
                                  Biochem, Mann

98. Nuclease, microccal   NBC, Schwarz, Sigma,
                                  Worthington

99. Nucleoside phosphory-   Boehringer
       lase, calf spleen

100. 5-Nucleotidase, venom   BDH, Sigma

101. Oxalate decarboxylase,   NBC, Schwarz, Sigma,
        C. velutipes   Worthington

102. Papain, papaya latex   BDH, Calbiochem,
                                  General Biochem, Mann,
                                  Merck, NBC, Pierce,
                                  Schwarz, Sigma, Wor-
                                  thington

103. Pectinase, mold   Calbiochem, General
                                  Biochem, Mann, NBC,
                                  Pierce, Sigma, Waller-
                                  stein

104. Pectin methyl esterase,   Mann, NBC, Sigma,
        tomato   Worthington

105. Penicillinase   Calbiochem, Mann,
                                  NBC, Pierce

106. Pepsin, hog stomach   BDH, Calbiochem,
                                  Cudahy, General Bio-
                                  chem, Mann, Merck, NBC,
                                  Pentex, Pfanstiehl,
                                  Pierce, Schwarz, Sigma,
                                  Worthington

107. Pepsinogen, hog stomach   General Biochem, Mann,
                                  NBC, Pentex, Pierce,
                                  Schwarz, Sigma, Wor-
                                  thington

108. Peptidase, hog intestine   Calbiochem, General Biochem, Mann, NBC, Pentex, Pierce, Sigma

      Klebsiella aerogenes   Pierce

109. Peroxidase, horseradish   BDH, Boehringer, Calbiochem, Mann, NBC, Pentex, Pierce, P-L Biochemicals, Schwarz, Seravac, Sigma, Worthington

110. Phenylalanine decarboxy-   Sigma
     lase, Strep.

111. Phosphatase, acid,   Boehringer, Calbiochem, potato   Mann, Sigma

     wheat germ   BDH, Calbiochem, Mann, Miles, Pierce, Schwarz, Sigma, Worthington

112. Phosphatase, alkaline,   BDH, Boehringer, Calbiochem, General Biochem, Mann, Miles, Pierce, P-L Biochemicals, Seravac, Sigma
     calf intestine

     chicken intestine   Mann, Worthington

     E. coli   Miles, Schwarz, Sigma, Worthington

     hog intestine   Sigma

113. Phosphodiesterase,   Mann, NBC, Pierce, bovine spleen   Schwarz, Sigma, Worthington

     venom   BDH, Boehringer, Calbiochem, NBC, Pierce, Sigma, Worthington

114. Phosphoglucomutase,   Boehringer, Calbiochem, rabbit muscle   Sigma

115. 6-Phosphogluconic dehy-   Boehringer, Calbiochem, drogenase, yeast   Sigma

116. Phosphoglucose iso-   Calbiochem, Sigma
     merase, rabbit muscle

     yeast   Boehringer, Calbiochem, Mann, Sigma

117. Phosphoglycerate mutase, rabbit muscle	Boehringer, Calbiochem, Sigma
118. 3-Phosphoglyceric phos- phokinase, yeast	Boehringer, Calbiochem, General Biochem, Mann, Sigma
119. Phospholipase (see Lecithinase)	
120. Phosphoriboisomerase, spinach	Pierce, Sigma
121. Phosphoribulokinase, spinach	Sigma
122. Phosphorylase, A and B, rabbit muscle	General Biochem, Mann, NBC, Schwarz, Sigma, Worthington
123. Phosphorotransacetylase, Cl. kluyveri	Boehringer, Sigma
124. Polynucleotide phosphory- lase, M. lysodeik- ticus	BDH, Calbiochem, Mann, Miles, P-L Biochem- icals
E. coli	General Biochem (Bio- polymers Div.)
125. Pronase	BDH, Calbiochem
126. Protease, bacterial	Calbiochem, Mann, Miles, Sigma, Wallerstein
bovine pancreas	NBC, Schwarz, Sigma, Worthington
fungal	Miles, Sigma, Waller- stein
papaya	Sigma
vegetable	Wallerstein
127. Proteinase, A. oryzae	P-L Biochemicals
128. Pyruvate decarboxylase, yeast	Boehringer
129. Pyruvate kinase, rabbit muscle	Boehringer, Calbiochem, General Biochem, Mann, NBC, Sigma

130. Rhodanese, beef liver   Sigma

131. Ribonuclease, pancreas   BDH, Boehringer, Cal-
                                     biochem, General Bio-
                                     chem, Miles, NBC,
                                     Pentex, Pierce, P-L
                                     Biochemicals, Schwarz,
                                     Seravac, Sigma, Wor-
                                     thington

     A -   Calbiochem, Mann,
                                    Miles, NBC, Pierce,
                                    P-L Biochemicals,
                                    Schwarz, Seravac, Sigma

     B -   NBC, Pierce, Schwarz,
                                    Sigma, Worthington

     D -   Calbiochem, Mann,
                                    Pierce, P-L Biochemicals,
                                    Seravac

     T_1 (A. oryzae)   Calbiochem, Mann, Miles,
                                    NBC, Schwarz, Sigma,
                                    Worthington

     Reduced (S-carboxy-   Mann, Sigma
          methylated)

     1-Carboxymethyl-   Mann, Sigma
          histidine-119

     Oxidized   Mann, Sigma

132. RNA Phosphorodi-   Miles
          esterase,
          M. lysodeikticus

133. RNA Polymerase, M.   Miles, Sigma
          lysodeikticus

     E. coli   General Biochem (Bio-
                                    polymers Div.)

134. Sorbitol dehydrogenase,   Boehringer
          sheep

135. Steroid dehydrogenase
          (See hydroxysteroid
          dehydrogenase)

136. Streptokinase   NBC

137. Sulfatase, Helix pomatia   Pierce, Sigma
      Limpets   Sigma

138. Tautomerase, beef kidney   Sigma
      pig kidney   Sigma

139. Thiamine pyrophospha-   Wallerstein
      tase, A. oryzae

140. Thrombokinase, rabbit   Pierce

141. Transfructosylase,   Pierce
      A. niger

142. Transglucosylase,   Pierce
      A. niger

143. Triosephosphate iso-   Boehringer, Calbiochem
      merase, rabbit
      muscle

144. Trypsin, pancreas   BDH, Boehringer, Cal-
      biochem, Cudahy, General
      Biochem, Mann, Merck,
      Miles, NBC, Pentex,
      Pfanstiehl, Pierce,
      P-L Biochemicals,
      Schwarz, Seravac, Sigma,
      Worthington

      acetylated   Mann, Pierce, Seravac,
      Sigma

145. Trypsinogen, pancreas   BDH, Boehringer, Cal-
      biochem, Mann, NBC,
      Pierce, P-L Biochem-
      icals, Seravac, Sigma,
      Worthington

146. Trypotophanase, E. coli   Sigma

147. Tyrosinase (Polyphenol   Mann, NBC, Pierce,
      oxidase), mushroom   Schwarz, Sigma, Wor-
      thington

148. Tyrosine decarboxylase,   NBC, Sigma, Worthing-
     APO Enzyme   ton

     Strep. faecalis   Calbiochem, Mann, NBC,
                                   Schwarz, Sigma, Wor-
                                   thington

149. Urease, jack bean   BDH, Calbiochem,
                                   General Biochem, Mann,
                                   Merck, NBC, Pierce,
                                   Schwarz, Sigma, Wor-
                                   thington

150. Uricase, bovine kidney   Mann, NBC, Schwarz,
                                   Sigma, Worthington

     C. utilis   Miles, Worthington

     hog   BDH, Boehringer, Cal-
                                   biochem, Mann, Pierce,
                                   Seravac, Sigma, Wor-
                                   thington

151. Uridine-5'-diphospho-   Sigma
         glucose dehydro-
         genase, bovine
         liver

152. Urokinase, human urine   NBC, Pierce

153. Xanthine Oxidase, milk   Boehringer, Mann, NBC,
                                   Schwarz, Sigma, Wor-
                                   thington

ADDRESSES OF COMPANIES SUPPLYING BIOCHEMICAL REAGENTS

Boehringer	Boehringer Mannheim Corp. 219 East 44th St. New York, N. Y. 10017
	C. F. Boehringer and Soehne, Mannheim, Germany
British Drug Houses	British Drug Houses, Ltd., Poole, England
	British Drug Houses Canada, Ltd., Barclay Avenue Queensway, Toronto, Canada
Brinkman (U. S. Distributor for E. Merck, Darmstadt, Germany)	Brinkman Instruments, Inc. Catiague Rd., Westbury, N. Y. 11590
Calbiochem	Calbiochem Co., P. O. Box 54282 Los Angeles, Calif. 90054
City Chemical	City Chemical Corp., 132 West 22nd St. New York, N. Y. 10011
Cudahy	Cudahy Labs 5014 South 33rd Omaha Nebraska 68107
Cutolo Calosi	Farmachimica Catolo Calosi, Naples, Italy
Cyclo Chemical	Cyclo Chemical Div. of Travenol Labs 1922 E. 64th Los Angeles, Calif. 90001

Fermco

Fermco Laboratories
Div. of Searle and Co.
P. O Box 5110
Chicago, Ill. 60680

Fisher

Fisher Scientific Co.
711 Forbes Ave.
Pittsburgh, Pa. 15219

Gallard-Schlesinger
(U. S. Dist. for Seravac Labs)

Gallard-Schlesinger
Chemical Mfg. Corp.
584 Mineola
Carle Place, Long Island
New York   11514

General Biochemicals

General Biochemicals
925 Laboratory Park
Chagrin Falls, Ohio
                    44022

Light

Koch-Light Laborato-
ries, Ltd.,
Colnbrook, Bucks
England

Mann

Mann Research Labora-
tories
136 Liberty
New York, N. Y. 10006

Merck

E. Merck AG
61 Darmstadt
Germany

Miles Labs

Miles Laboratories, Inc.
Research Products Div.
Elkhart, Indiana 46514

Mycofarm-Delft

Mycofarm-Delft
Div. of Royal Nether-
lands Fermentation
Industries,
Delft, Holland

NBC

Nutritional Biochemicals
Corp.,
26201 Miles Road
Cleveland, Ohio 44128

Pentex

Pentex, Inc.
P. O. Box 272
Kankakee, Ill. 60901

Pfanstiehl

Pfanstiehl Labs, Inc.,
1219 Glen Rock Ave.
Waukegan, Ill. 60085

Phylab

Phylab
Div. of Physicians and
Hospitals Supply Co.
1400 Harman Place
Minneapolis, Minn.
55403

Pierce

Pierce Chemical Co.
P. O. Box 117
Rockford, Ill. 61105

P-L Biochemicals

P-L Biochemicals, Inc
1037 W. McKinley Ave.
Milwaukee, Wis. 53205
(Formerly Pabst Labs)

Regis

Regis Chemical Co.
1101 N. Franklin
Chicago, Ill. 60610

Schwarz

Schwarz Bio Research
Mountain View Avenue
Orangeburg, N.Y. 10962

Seravac

Seravac Laboratories
Maidenhead, England
and Capetown,
S. Africa
(U. S. Distributor -
Gallard-Schlesinger)

Sigma

Sigma Chemical Co.
3500 De Kalb St.
St. Louis, Mo. 63118

Wallerstein

Wallerstein Co.
6301 Lincoln Ave.
Morton Grove, Ill. 60053

Worthington

Worthington Biochemical
Corp. ,
Freehold, N. J. 07728

AUTHOR INDEX

Numbers in parentheses are reference numbers and indicate that an author's work is referred to although his name is not cited in the text. Numbers underlined show the page on which the complete reference is listed.

315

E

Edel, F., 58, 106
Edelstein, M., 62, 106
Edson, N. L., 161(110), 173
Egami, F., 166(137), 175
Eggstein, M., 161, 173
Ehrlich, G., 166, 174
Eik-Nes, K. B., 165, 174
Elevitch, F. R., 65(59), 72, 98(154,155), 107, 108, 111, 119, 169
Elhilai, M. M., 102, 112
Ellman, G., 47, 105
Emmens, C. W., 63(50), 107, 153(92), 172
Emory, E., 97(151), 111
Engelhard, W. E, 167(139), 175
Enterline, M. L., 72(75), 108
Epstein, J., 202(32), 221, 232
Erlanger, B. F., 19(18), 24, 58, 106
Espersen, G., 104(179), 112

F

Fallscheer, H. O., 202(29), 218, 223(29), 232
Faway, E., 168, 175
Faway, G., 166, 168, 174, 175
Feichtmeir, T. V., 72(75), 108
Fellig, J., 202(64), 230(64), 234
Fernley, H. N., 72, 74(77), 108
Ferrante, N., 63(47), 107, 153(93), 172
Field, G. F., 1(1), 23, 179(11), 182(11), 183(11), 195
Fink, G., 162, 174
Fischer, A., 229(62), 234
Fischer, O., 71(71), 108

Fischerova-Bergerova, V., 36(19), 41
Fishman, W. H., 62(39), 106
Folin, O., 129(54), 171
Fouts, J. R., 127(40, 41), 170
Fowler, K. S., 202(40), 223, 233
Frankel, S., 104(180), 112
Franz, P., 202(47), 224, 233
Freed, S., 58, 106
Freiden, C., 99(160), 111
Frings, C., 86(113), 109, 158, 173, 271(14), 276(14), 281(20), 294
Fritz, H., 260(29,30), 264
Fuchs, E., 76(80), 108
Fungmann, U., 141(81), 172
Fyowa, 86(119), 110, 118(15), 169

G

Gadaleta, M. N., 36(22), 41
Gale, E. F., 114(1), 135(72, 73), 137(72, 73), 168, 172
Gall, E. G., 90(134), 110
Galstatter, J. H., 202(36), 222, 233
Gamson, R. M., 45, 105, 202(34), 222, 232
Gatfield, P. D., 125(35), 170
Gaubert, J. P., 160(108), 173
Gavard, R., 160(108), 173
Gebbing, H., 161, 174
Gennaro, W., 103, 112
Gerez, C., 166, 174

Hall, D. A., 32, <u>41</u>, 83,
  <u>109</u>, 116, <u>169</u>
Hall, S. A., 2<u>02</u>(28), 218,
  221(28), 222, <u>232</u>
Hankinson, D. J., 202(<u>49</u>),
  224, <u>233</u>
Hansen, A. P., 103(<u>175</u>), <u>112</u>
Hansen, P. F., 104(<u>178</u>),
  <u>112</u>
Happold, F. C. 160(107),
  <u>173</u>
Hartree, E. F., 83(98),
  84(105), <u>109</u>, 116(6),
  <u>169</u>
Hausman, T. V., 71(69),
  <u>108</u>
Hayakawa, T., 179, 183(9),
  184, <u>195</u>
Haynes, J., 72, <u>108</u>
Hehler, A. M. 10<u>1</u>(165),
  <u>112</u>
Helger, R., 71(69),
  <u>108</u>
Helmke, E., 119, <u>169</u>
Henley, K S., 10<u>4</u>(176),
  <u>112</u>
Heppel, L. A., 90(136),
  <u>110</u>
Herrlinger, F., 88(124),
  <u>110</u>
Herrmann, H., 229(62),
  <u>234</u>
Hersch, R. T., 76(80),
  <u>108</u>
Heyn, A., 55, <u>105</u>
Hick, F. L., 10<u>0</u>(162),
  <u>112</u>
Hicks, G. P., 10(12), <u>24</u>, 83,
  98, <u>109</u>, <u>111</u>, 118, <u>169</u>,
  177(<u>3</u>, 4), 178(3), <u>179</u>,
  180(4), 183(3, 4), <u>195</u>,
  200(12), 201(12), 20<u>7</u>,
  208(12), 209(12),
  210(12), <u>232</u>, 246, 248,
  249(7), 2<u>50</u>(7), 251,
  25<u>2</u>(7), 253, 254,
  255(9), 256(9), 257(9),
  262, <u>263</u>, 265, 274,
  275(1<u>7</u>), 276, 277(17),
  <u>293</u>, <u>294</u>

Hieserman, J., 78, <u>109</u>,
  132, 133(71), <u>135</u>,
  136(71), <u>171</u>, 186,
  <u>195</u>
Hill, B., 97(151), <u>111</u>
Hill, J. B., 289(28),
  <u>294</u>
Hill, R. M., 167(139),
  <u>175</u>
Hobom, G., 202(65),
  230, <u>234</u>
Hodapp, P., 119(26),
  <u>169</u>
Hoffman, P., 63(45),
  <u>106</u>
Hofman, G., 60(34),
  <u>106</u>
Hofmann E., 60(34), <u>106</u>
Hogness, T. R., 184(<u>18</u>),
  185(18), <u>195</u>
Hohorst, H. J.,
  192(36), <u>196</u>
Holldorf, A., 160(106),
  <u>173</u>
Hollister, L., 119,
  <u>169</u>
Holmsen, H., 185(37),
  193(37), <u>196</u>
Holmsen, I., 18<u>5</u>(37),
  193(37), <u>196</u>
Holmstedt, B.,
  128(44, 45), <u>170</u>
Holzer, H., 160(10<u>6</u>),
  <u>173</u>
Horecker, B. L., 90(136),
  <u>110</u>, 115(2),
  <u>122</u>(30), <u>168</u>, <u>170</u>,
  184(18), <u>185</u>(18),
  <u>195</u>
Horn, H. D., 202(59),
  229(59, 60), <u>234</u>
Horwitt, M. K., 2<u>29</u>(61),
  <u>234</u>
Hu, A., 126(36), <u>170</u>
Hübener, H. J., 16<u>4</u>(128),
  <u>174</u>
Huff, J. W., 96, <u>111</u>
Huggins, C., 71, <u>107</u>
Humme, B., 58(28),
  <u>106</u>

SUBJECT INDEX

with tyrosine decarboxy-
lase, 137
Tyrosine decarboxylase,
source of, 311
use in assay of L-tyro-
sine, 137
Tubular platinum electrode,
use in assay of glucose,
278-280

U

Urastat, 130
Urea, assay of with urease,
129-131,256
colorimetric methods,
129-130
electrochemical
methods, 130-131
fluorometric methods,
130
automated assay of,
286,287,291
specific electrode for,
256
Urease, assay of, 76-77
immobilization of, 256
source of, 311
use in assay of,
cadmium, 200,215,216
cobalt, 200,215,216
copper, 201,215,216
lead, 201,215
manganese, 201,215
mercury, 201,215,216
nickel, 201,215,216
silver, 200,215,216
urea, 129-131,256
zinc, 201,215
Uric acid, automated assay
of, 289
Uricase, source of, 311
use in assay of uric
acid, 289
Uridine-5'-diphospho-
glucose dehydrogenase,
source of, 311
Urokinase, source of, 311

V

D-valine, assay of,
132,134
Variable time, methods
of assay of enzyme
kinetics, 268-272

X

Xanthine, assay of,
162
spectra of, 163
Xanthine oxidase, assay
of, 89-92
electrochemical
methods, 90,91
fluorometric
methods, 90,92
radiochemical
methods, 92
spectrophotometric
methods, 89-90
source of, 311
use in assay of, hypo-
xanthine and
xanthine, 162,163
o-iodisobenzoic
acid and p-
chloromercuri-
benzoate, 202,
229
mercury, 201,216
other aldehydes,
162
silver, 200,216
Xylitol, assay of,
161

Z

Zinc, assay of, by
activation of iso-
citrate dehydroge-
nase, 179,180,183
by inhibition of
urease, 201,215
Zymat 340 analyzer, 283,
285

OTHER TITLES IN THE SERIES IN ANALYTICAL CHEMISTRY